The World After Cheap Oil

Substantial evidence suggests that we are currently living at the peak of oil production with few prospects for cheap oil ever returning. Yet the media, politicians, and the general public have hardly started to talk about what this means. Oil literally runs our societies, from transportation to food production to economic activity. Without oil, everything stops. There are powerful arguments that if we fail to increase oil production, we will also fail to grow our economy as a whole. For oil-importing western nations the news is bleak; higher oil prices seem to put a glass ceiling on their economic growth, making current debt problems worse no matter what monetary and economic policies we might choose.

The World After Cheap Oil offers a thorough package of information about oil; its uses and its role in our society's important sectors. It presents the most prominent substitutes and alternatives, and their limits and promises. It also delves deep into the many risks, problems, and mechanisms that can make the world after cheap oil a much more unstable place for nations and humanity as a whole. The book also explains why there has been so little public debate on the subject, and what the future might look like after oil production starts its final, terminal decline.

Rauli Partanen – President and founder of Peak Oil Finland (a non-profit organization), independent writer, lecturer, and consultant on the environment, energy, society, the economy and their interrelations. B.B.A.

Harri Paloheimo – Founder of Peak Oil Finland; CEO and co-founder of Coreorient Ltd. developing crowdsourced deliveries; MSc; PhD student of Aalto University on transport energy efficiency; consultant for companies, cities, and the Finnish Parliament in the areas of energy and transport.

Heikki Waris – Founder of Peak Oil Finland; co-founder of Coreorient Ltd. developing crowdsourced deliveries; MSc; consultant for companies, cities, and the Finnish Parliament in the areas of energy and transport.

The World After Cheap Oil

Rauli Partanen, Harri Paloheimo
and Heikki Waris

Routledge
Taylor & Francis Group
LONDON AND NEW YORK

First published 2015
by Routledge
2 Park Square, Milton Park, Abingdon, Oxfordshire OX14 4RN

and by Routledge
711 Third Avenue, New York, NY 10017

First issued in hardback 2015

Routledge is an imprint of the Taylor & Francis Group, an informa business

British Library Cataloguing-in-Publication Data
A catalogue record for this book is available from the British Library

Library of Congress Cataloging-in-Publication Data
Partanen, Rauli.
The world after cheap oil/Rauli Partanen, Harri Paloheimo and Heikki Waris.
pages cm
Includes bibliographical references and index.
1. Petroleum reserves. 2. Petroleum products. 3. Petroleum industry and trade. 4. Energy consumption. 5. Energy policy.
I. Title.
HD9560.5.P28 2015
338.2'7282—dc23
2014018680

ISBN 13: 978-1-138-13236-8 (hbk)
ISBN 13: 978-1-138-80637-5 (pbk)

Typeset in Sabon
by Florence Production Ltd, Stoodleigh, UK

Contents

List of illustrations vii
Foreword ix
Preface xi
Acknowledgements xiii
Terms and abbreviations xv
Introduction: without oil, there is no industrialized society xix

1 Oil 1

2 The end of cheap oil 26

3 What instead of oil? 73

4 Oil in the modern, industrialized society 128

5 World after cheap oil 191

Appendices 235
Index 243

Illustrations

Figures

1.1 Food prices and oil prices, 2000–2010 5
1.2 World oil supply growth vs world GDP growth 7
1.3 The development of IEA predictions for world oil demand in 2030 or 2035 20
2.1 Net energy for society drops rapidly when EROEI goes under 8:1 29
2.2 Crude oil quantity extracted and price 40
2.3 Egypt: oil – 2012 imports increased by 282 percent 43
4.1 Trends in oil consumption of some European Union member countries, 1970–2013 131
4.2 Historical oil consumption for the US, Canada, Russia, China, and Japan, 1970–2013 133

Tables

1.1 Growth in different categories between 2005 and 2012 15
4.1 Top 12 producers in 2012 129
4.2 Top 12 exporters in 2012 130
4.3 Top 12 importers in 2012 130
4.4 Russian oil reserves by different sources 136
4.5 Paved roads and railways in some countries and areas (2008) 145
4.6 Primary energy sources in EU 163

Foreword

In March 1998, *Scientific American* published a controversial, groundbreaking article by two distinguished petroleum geologists, Colin Campbell and Jean Laherrère. Many readers thought the article's title, "The End of Cheap Oil," to be unhinged from reality. After all, petroleum was then selling for $12 per barrel, the lowest price ever in inflation-adjusted terms. Campbell and Laherrère explained that this was a temporary fools' paradise that would inevitably end as a result of two long-term trends: falling average annual rates of oil discovery, and rising oil consumption. Taking these trends into account, a peak in world oil production should be foreseeable: "within the next decade, the supply of conventional oil will be unable to keep up with demand."

The events of the past 16 years have thoroughly vindicated "The End of Cheap Oil." Petroleum prices have settled at a "new normal" of over $100 per barrel and the world rate of production of conventional oil (not counting biofuels, refinery gains, natural gas liquids, and other combustible liquids that cannot be sold as crude oil) stopped growing in 2005.

The petroleum industry, meanwhile, has dramatically increased spending on exploration and production – and has seen rapidly diminishing returns from its investments. Between 1998 and 2005, the industry spent $1.5 trillion on exploration and production, yielding 8.6 million barrels per day in additional world oil production. Between 2005 and 2013, the industry invested $4 trillion in exploration and production (E&P), yet this more-than-doubled investment produced only 4 mbpd (million barrels per day) in added production – and all of that increase came from problematic, unconventional sources.

Nevertheless, oil industry spokespeople have consistently and vehemently derided all discussion of "peak oil" as rubbish. Their only significant source of evidence for continued bullishness on petroleum supplies is recent success in production of oil from low-permeability rock formations in North Dakota and south Texas, using the technologies of hydraulic fracturing ("fracking") and horizontal drilling. True, U.S. oil production has increased by over 2.5 mbpd as a result of the fracking frenzy. However, the "tight oil" boom would never have happened without high oil prices (resulting from declines in production growth from conventional fields). Moreover, there is good reason to believe the boom will be very brief and will not be replicated to any significant degree in other parts of the world (for details, please see the Post Carbon Institute study, "Drill, Baby, Drill," at www.shalebubble.org).

This is not to say the world is running out of oil. Far from it: there are many more petroleum (or petroleum-like) deposits still lurking in the earth's crust than have been targeted by drillers so far. Our problem is that the industry has understandably found and produced the best oil first, leaving the more expensive, harder-to-access, and more environmentally hazardous oil for later.

It is later. And the oil that is left today is quite different from the oil that fueled the industrial miracle of the twentieth century. The implications for our transport, food, and economic systems could hardly be more serious.

Rauli Partanen, Harri Paloheimo, and Heikki Waris have given us more than a peek into the world after cheap oil. This book is an excellent introduction to, and overview of, what has grown to be an intricate, controversial, ever-developing, and profoundly important field of study – the likely future trajectory of world oil production, and the implications of that trajectory for an overwhelmingly petroleum-dependent world. The authors are to be commended for covering nearly every relevant topic within this field, and doing so in a way that is clear and fair.

Reader be forewarned: a heightened appreciation and understanding of our society's perilous dependency on this depleting, non-renewable, and climate-changing resource has been known to result in significant changes in behavior (reduced consumption) and attitude (more skepticism toward reassurances by government and industry leaders that business-as-usual can be maintained). If you are already familiar with the peak oil literature, this book will provide you with a comprehensive, up-to-date summary you can use in educating friends and relatives. If you are new to this discussion, prepare to have your world shaken.

Richard Heinberg
Senior Fellow, Post Carbon Institute
Author, *The End of Growth: Adapting to Our New Economic Reality*

Preface

It has been a work of deep contradiction to write this book. After all, the picture we ended up with was not something that we would wish to actually become a reality. So while we are proud of, and stand by, the work we have done, we also hope that it will somehow prove to be not needed. Maybe there will be sufficient action to prevent the darker possible scenarios becoming a reality. Maybe some of that action will even be due to this book, as well as many other great books and articles written on this subject. In a sense, the aim of this book is to make itself useless.

It is also possible that the risks, predictions, and trends presented in this book will prove to be much more pessimistic than reality. This would be great, as while the authors would no doubt earn a reputation as dystopian visionaries, the rest of humanity would do much better.

So while we fear that we have been relatively correct, or even too optimistic in many things, we hope from the bottom of our hearts that the worrisome message of this book will prove to be darker and more pessimistic compared with humanity's actual future. One way or another.

Acknowledgements

Rauli Partanen would like to thank all the people who believed that this book could be written, and those who did not believe it, but held their tongue. The Finnish Association for Non-fiction Writers deserves my thanks for their grant on the book *Finland After Oil*, which eventually led to this book; as does my wife, for providing for our family while I was writing; and our children Niilo and Eino for providing sufficient distraction and reminding me of what is actually important in life. Harri and Heikki, my co-authors, colleagues and, because and despite of this book project, dear friends, also deserve my gratitude. Without their invaluable input and knowledge, this book would be a mere shadow of what it is. Celia Peterson deserves a big Thank You for working as our copy-editor and for proofreading my non-native English. A further thank you goes to our publisher Routledge and our editor Lisa Reading (not forgetting Trudy Varcianna) for putting up with us.

Harri Paloheimo would like to thank all those rare individuals in different organizations who have realized the risks involved in peak oil, and have given me strength to go on. This book would never have seen the light of day were it not for Rauli's perseverance, and he also carries the honor of molding our message into a readable and understandable format. I would also like to thank Heikki for our years of trying to find ways to mitigate the oil dependency of societies, and for his exceptional courage in pushing onwards while surrounded by skeptics. I would especially like to thank my wife Aura for her unwavering support, and our children Pinja and Talvi for being the hope in our future that gave me the reason to do any of this.

Heikki Waris would like to thank all those people who contributed their knowledge and analytical skills to discuss and educate the rest of us on the many topics that receive insufficient attention in the mainstream media. It takes a special kind of courage to go against the flow. I also want to thank you, dear reader. You are the best hope for getting our societies to prepare for what is coming. I would like to thank Harri for his endless energy that supports those around him. A special thank you goes to Rauli, who wrote and rewrote the book again and again, day after day, chapter by chapter, until it was ready. And finally I would like to thank Katri and Lea for being there and showing that happiness is an attitude separate from the troubled world around us.

Terms and abbreviations

Abbreviations

ASPO International	Association for the Study of Peak Oil and Gas
BITRE	(Australian) Bureau of Infrastructure, Transport and Regional Economics
BRICS (countries)	Brazil, Russia, India, China, South Africa
BtL (process)	biomass to liquid
BWR	boiling water reactor
CAFE (standard)	corporate average fuel economy
CERA	Cambridge Energy Research Associates
CNG	compressed natural gas
CCS	carbon capture and storage, or carbon capture and sequestration
CtL (process)	coal to liquid
DME	dimethyl ether
DCL	direct coal liquefaction
E&P	exploration and production
EIA	(U.S.) Energy Information Administration
EMROI	energy money return on investment
EOR	enhanced oil recovery
EPA	Environmental Protection Agency
EROEI (also EROI)	energy returned on energy invested
ETF	exchange-traded fund
EV	electric vehicle
GDP	gross domestic product
GE	genetic engineering
GMO	genetically modified organism
GtL (process)	gas to liquid
ICE	internal combustion engine
ICL	indirect coal liquefaction
IEA	International Energy Agency
IFR	integral fast reactor
IMF	international monetary fund

IPCC	Intergovernmental Panel on Climate Change
ITPOES	(British) Industry Taskforce on Peak Oil and Energy Security
LNG	liquid (liquefied) natural gas
LPG	liquid petroleum gas
LTO	light tight oil
MtG (process)	methanol to gasoline
NER (fund)	new entrants' reserve fund
NG	natural gas
NGL	natural gas liquid
NGPL	natural gas plant liquids
NGOs	non-governmental organizations
OECD	Organization for Economic Cooperation and Development
OPEC	Organization of the Petroleum Exporting Countries
PPP	purchasing power parity
PWR	pressurized water reactor
ROI	return on investment
SAGD	steam assisted gravity drainage
SEC	Securities and Exchange Commission
SUV	Sport Utility Vehicle
USGS	United States Geological Survey
WEO	(IEA) *World Energy Outlook*

Units

mcf	thousand cubic feet (of gas)
mmcf	million cubic feet
bcf	billion cubic feet
tcf	trillion cubic feet
qcf	quadrillion cubic feet
bpd	barrels per day
mbpd	million barrels per day
Gb	gigabarrels (or billion barrels)
kW (power)	kilowatt
MW (power)	megawatt
GW (power)	gigawatt
kWh (energy)	kilowatt-hours
MWh (energy)	megawatt hours (= 1,000 KWh)
GWh (energy)	gigawatt hours (= 1,000 MWh)
TWh (energy)	terawatt hours (= 1,000 GWh)
GJ (energy)	gigajoule
tkm	ton-kilometers
ppm	parts per million

Terms and concepts

The following list outlines some of the key terms and concepts used throughout the book that may be unfamiliar to some readers. The terms mostly referred to are those used by the International Energy Agency:

Conventional oil	Includes crude oil, natural gas liquids, and condensates.
Unconventional oil	Includes oilshale, oil sands, extra heavy oil and bitumen, synthetic oil, and oil liquefied from natural gas or coal.
All liquid fuels	Includes biofuels and other possible liquid fuels such as refinery gains which are increases in volume due to refining.

Some institutions and organizations count all the above as oil and some do not, which helps to explain the differences in figures given for oil produced and oil products consumed:

Peak Oil	A term, which refers to the "maximum rate of the production of oil in any area under consideration, recognising that it is a finite natural resource, subject to depletion" (Colin Campbell, ASPO International). The Peak in conventional oil was most likely between 2006 and 2008, according to the IEA.
Resources	Total amount of original oil in place (not to be confused with *reserves*).
Reserves	Estimated amount that can be produced with current prices and technology: • 1P (or P90) oil reserves have a 90 percent probability of being produced at current prices and with current technology (1P = *proven*) • 2P reserves (or P50) probability is 50 percent (2P = *proven* + *probable*) • 3P reserves (or P10) probability is 10 percent (3P = *proven* + *probable* + *possible*).
R/P (reserves to production rate)	Dividing the proven reserves of an oilfield by its current production gives the number of years that can be produced from the field at current speed. From the perspective of peak oil it is not very usable, as it does not account for the fact that production starts to decline long before the oil is running out.
Primary energy	The fuel or energy available for use before it is transformed into a more usable form, such as burning coal to make electricity (coal being the

	primary energy and electricity being the secondary energy).
Secondary energy	Electricity, liquid fuels, any heat that is not wasted.
Recovery factor	Refers to the amount of oil (usually 20–60 percent) in a conventional oilfield that can be extracted.
Exergy	The energy that can actually be used for some practical purpose. 1 kWh of liquid fuel has more exergy than 1 kWh of ambient heat, since this cannot be used to do much useful work. The share of primary energy that can be captured and used as exergy is defined by the efficiency of our energy use.
Price of oil	Normally refers to Brent oil prices, U.S. dollars per barrel.
Units	A barrel is the common unit of oil. Globally, the normal unit is mbpd, or million barrels per day.

Introduction

Without oil, there is no industrialized society

We live in a world that is completely hooked to oil. Without it nothing moves. The economy collapses. Food production collapses. Other forms of energy production collapse. The industrialized society in its present form collapses.

While the oil is not running out, the era of cheap oil is over. In addition to the laws of geology this is due to a few other causes as explained in this book. However, they all lead to the conclusion that there is no going back to the era of cheap oil – and cheap oil cannot be replaced by expensive oil.

Every day around 90 million barrels of oil are consumed worldwide. Of this, 13 million barrels are consumed in the European Union, 18.5 million barrels in the United States, 10 million barrels in China and only 3.5 million barrels in Africa. The EU produces only 1.5 million barrels per day (mbpd), so it imports more than 11 mbpd. At per capita level this is in the same ballpark as the U.S. oil imports. With current oil prices, the speed with which we are transferring wealth to oil exporting countries is staggering.

Western citizens use their share of the global oil several times over. More than half of the oil we use globally goes to transportation. The rest is divided between heating, construction, agriculture and forestry, industrial energy, raw materials, lubricants, and electricity production. Once the oil has been burned, it can never be recovered.

The growing oil production is one of the most important factors behind the rise in the standard of living. The historical trend shows that the economy grows hand in hand with oil consumption. But what happens when oil production peaks and begins to decline? We can expect economic hardships, resource wars, rising oil prices and a tighter geopolitical atmosphere. Looking at the news, you can notice that all of these are, at varying degrees, already in progress; and so is the end of cheap oil, as new oil is becoming more and more expensive to find and to extract.

Conventional oil production has practically stagnated for almost a decade. We find ourselves scraping various substitutes wherever we happen to find them: tight oil, oil sands, oil shale, biofuels, liquefied coal, liquefied petroleum gases, and so on. Anything goes as long as it burns. Yet these substitutes cannot come near to replacing the volume of conventional crude oil we are using.

At a minimum, a temporary drop in the availability of oil – a supply crunch – is ahead of us, estimated for around the middle of this decade, and it might be soon followed by a permanent downturn in the production of oil. This does not depend so much on the remaining oil reserves. The timing of the phenomenon can be calculated based on the expected production of oil projects that are currently underway and the declines in the production of existing oil fields. Bringing new oil projects into a production stage usually takes several years, even a decade. The exception is the recent tight oil production boom in the United States, which will likely face worsening problems and a peak production of its own in the near future.

At the same time, China and some other countries are rapidly increasing their consumption. This means that someone else has to sacrifice part of their share. In recent years that someone else has been largely OECD countries, whose economies are experiencing severe problems. These problems have been caused by excessive indebtedness, but also by expensive oil. Expensive oil has destroyed the demand for oil. This has also destroyed all the economic activity that oil could have made possible, directly and indirectly. Going into debt is a way to borrow from the future, and it seems that the future has finally arrived for many.

What, then, does the world look like through peak oil? The tightening oil markets create a fertile ground for a variety of black swans – unexpected events with major consequences. The shortfall of a few percent in global production levels caused by the civil war in Libya raised the price to a level that crushed the budding economic growth in Europe. Indicative of the scarcity of oil is that mere rumors of oil supply disruptions increase prices, and oil companies have to search for the next field from ever more difficult and accident-prone areas. There are no guarantees that the oil market remains open in the future. Many countries are already seeking to ensure their access to oil through bilateral agreements. Oil exporting countries are enjoying unprecedented prosperity, which the oil importing countries need to take from the pockets of their citizens.

How dependent are we on oil? And how well have the peaking of oil production and its consequences been taken into account in our national strategies and politics? European food production is dependent on imported energy and feedstock for fertilizers and different pesticides, and we heat our homes largely with Russian natural gas. Many cities in the U.S. have been built solely on the assumption of the passenger car, making them all but useless without it. In the background, hundreds of millions in China and India are planning to buy their first car.

Our retirement funds are entirely dependent on strong and sustained economic growth. With a rapidly aging population in Europe, increasing indebtedness and worsening economic outlook, these funds might soon be on the brink of a collapse, sowing distrust and bad blood between generations and social classes. Expensive and scarce oil could be the final nail in the coffin of sustained economic growth for the time being, leaving politicians in the unpleasant and unforgiving position of sharing the fruits of austerity instead of growth. So far, austerity has mainly gone to the common people, while the remaining fruits have been poured into the financial sector and the very wealthy. This has been done by making profits private but socializing losses to the public. Economic growth, where there has been any, has largely been the result of blowing up asset bubbles and growing the immaterial financial sector. But our society, the one that all of this is built on, still runs on oil.

What, then, is oil, and why is it so important?

1 Oil

What is oil?

For a lot of people, oil is a lubricant that gets changed when you take your car for its annual check-up. For others, the thought of oil brings up images of cooking and olive oil. Yet others also know that plastics are made of oil. It is a surprise for quite a few people that in addition to lubricants and plastics, diesel and gasoline are also made of crude oil. Another popular image might be the "nodding donkeys" from the television series *Dallas*. For the authors, as for many readers, oil used to be something vaguely abstract; something not very important, something in the background, that didn't really concern us. Oil, and the host of products made from it, has almost always been abundantly available. However, the truth about oil is a lot more complex than we think.

This book is mainly about oil as a fossil fuel. Mineral oil, petroleum, liquid fuel . . . it has many names. What we mean by oil in this book is one of the basic necessities of a modern industrialized society.

Oil is a hydrocarbon. It is densely packed energy in a convenient form that has a multitude of important uses. In its current volume, it is not replaceable by anything else. Oil is also the most important strategic resource in the world. What that means is that we don't have any other natural resource that would be more important. In this book we will explain why that is.

The common history of mankind and oil stretches for thousands of years. Mineral oil has been used in small quantities for this or that purpose throughout mankind's history. The birthplace of the modern oil industry is commonly considered to be Titusville, Pennsylvania. There Edwin Drake drilled his first modern oil well in 1859. However, oil had been produced in Europe and Russia for several decades before.[1] It was in the United States that the real oil rush was born, where it was distilled into kerosene to use in lighting. Oil was also used as a lubricant for machinery in the growing industrial manufacturing sector. Gasoline, as it happens, was an unwanted by-product of refining at the time due to its volatility. It was promptly disposed of by pouring it into the local rivers and waterways. Later, gasoline found its use in the modern internal combustion engine, and it is currently the most valuable product distilled from crude oil in refineries.

In the early twentieth century the United States overtook Russia as the biggest oil producer in the world. In the 1960s, mankind experienced its first peak as discoveries of oil spiked. The use of oil in the rapidly modernizing and industrializing nations grew rapidly as more and more uses were found for oil and its multitude of refined products.

Currently mankind is using over 90 million barrels per day (mbpd) of petroleum products. This amount is mind-boggling and hard to imagine. It is over 14 billion liters every day, or over 1,000 barrels (1 barrel equals 159 liters) per second. For each human, the average use is around 2 liters per day. Europeans use almost three times the average, and North Americans twice that amount. The energy content of crude oil is around 10 kilowatt-hours (kWh) per liter. For the average human, the daily physical work done is around 1 kWh. In using oil, we get the energy of around 10 "energy slaves" per each daily liter consumed. For the current price of oil, the hourly wage would be around 1 cent. Oil is, depending on your point of view, expensive or dirt cheap.

The birth of oil

According to the scientific consensus,[2] oil comes from the sunlight stored by algae and other biomass over the course of millions of years. This biomass first ended up in circumstances that prevented it being eaten by bacteria and suchlike, for example, at the bottom of a deep ocean. From there it eventually ended up in what is called the oil window – between 2 and 5 kilometers under the earth. There, given the proper pressure and temperature (between 74°C and 140°C), the biomass transformed into oil over millions of years. The biomass that ended up in deeper conditions was formed into natural gas. The biomass that ended up too near to the surface of the earth's crust was left in a sort of "raw" state, and is called kerogen. Contrary to popular belief, the oil in the ground is not formed in huge underground lakes, but rather is contained in pores of various sizes in the ground. The size and interconnectedness of these pores defines the permeability of the oil – or how easy or hard it is for the oil to move inside the rocks. This is one of the defining factors in how easy it is to produce oil from a given oilfield. Since oil and gas are lighter than other substances below ground, they tend to migrate up. So the further condition for a productive oilfield is a proper "cap" that will block the oil from migrating up. In addition, there needs to be a geological "trap" for the oil and gas to stay below the cap.

Not all oil is created equal, and a barrel of oil can contain thousands of different hydrocarbon molecules. Oil is defined by its lightness/heaviness, sourness/sweetness and various other measurements. Light and sweet crude oil is easily transported through pipelines and is cheap to refine into different products. It usually has a low content of sulfur, heavy metals, and other toxics. Heavy and sour oil on the other hand is much more expensive to refine, and it requires special equipment not found in all refineries.

The refined products can be crudely divided into four categories. These are: petroleum gases, light ends, medium distillates and heavy ends.

Petroleum gases (1–4 carbon atoms) include methane, ethane, propane, and butane. Methane is the main component in what is called natural gas. It is widely used in heating, cooking, and electricity production. It is also used as a source of hydrogen by the fertilizer industry[3] and for oil refining in processes such as hydrocracking. Ethane is widely used to make petrochemicals and plastics. Propane and butane can be used in barbequing and in cigarette lighters.

Light ends (5–11 carbon atoms) include, for example, gasoline and naphtha. Naphtha is used as feedstock for petrochemicals, plastics, and solvents. Gasoline is the most common transportation fuel, and also the most valuable refining product.

Medium distillates (11–18 carbon atoms) include kerosene used in jet fuel, lighting, cooking and heating. This group also includes fuel oils that are used as diesel fuel and home heating oil. Diesel engines used to be heavier and more expensive than the gasoline engine, and were therefore mainly used in heavy machinery, such as trains. With time and development, the use of diesel oil has also become more common in passenger cars, especially in Europe.

Heavy ends (18–50+ carbon atoms) include lubricating oils and residual fuel oil, which can be used as marine shipping fuel, industrial fuels, and for electricity generation. After the number of carbon atoms reaches above 25, the products tend to be solid instead of liquid. These are greases and waxes, bitumen (made into asphalt and roofing materials) and coke (used in steel production).

Different oil products are separated by distilling, or heating the crude oil to different temperatures, with the maximum around 400°C. When other measures, such as pressure and the addition of different catalysts, are applied, the refinery can increase the amount of different end products. Heavier molecules can be cracked into shorter ones, and short molecules can be combined to make longer ones. With these methods, the refinery can raise the share of gasoline products from the normal 20 percent up to 50 percent.[4]

There are differences between refineries. Not all refineries are equipped to handle heavy or sour oil. Generally, refineries that can refine heavy, high-sulfur oil, can also handle lighter oil, but not the other way around. In the future, a refinery's ability to handle lower quality oil will be a defining factor in its competitiveness in a tightening market.

Since refining is a business of thin margins, the capacities need to be kept high. This means that there cannot be much overcapacity in the world at any given moment. The profitability of refineries has been suffering recently because of the recession, additional investments demanded by the rise of biofuels, and global political uncertainty. When the production of oil eventually starts to fall, there will not be enough for all the refineries to refine.

Oil in our daily life

Few of us can imagine how tightly entangled our modern lives are in oil. Most know that plastics are made of petroleum, and, when you look around, you see that plastics are everywhere. In addition, the mark of oil can be seen in practically all consumer goods in either the production or the transport, or both. Oil is everywhere.

Often oil can be replaced with other materials, but rarely as cheaply and maintaining the same quality. Plastic is an excellent material for packaging. It helps to avoid losses due to spoiling in foodstuffs, for example. One can use natural materials for insulation, but the proper knowhow on their uses is not very common and building companies currently do not have many incentives to use them. While winter clothes can be made with fur, cotton or wool, these are also not without their problems. Synthetic fibers have a lot of good qualities that we have grown used to.

Plastics are everywhere, and they are everywhere for a good reason. As a material plastic is flexible, easily modeled, relatively durable, and, most importantly, cheap compared with other similar materials. While it is true that mankind has lived most of its existence without plastics, they are essential for anything resembling our modern way of life.

The single biggest user of oil is transportation, since over 90 percent of our transportation is fueled with oil. There are around 1 billion vehicles with internal combustion engines (ICE) that use liquid fuels such as diesel or gasoline as their main energy source. If oil were taken away from us today, the world would come to a halt within a few days. Stores would be empty in a few days as the panic spread. Stocks of consumer goods are kept low in order to minimize the need for capital and expenses, but, without oil, new goods would no longer arrive at stores even if they could somehow still be manufactured.

Even our most basic energy source, food, is nowadays made using petroleum and natural gas. For each calorie eaten in the developed world, 5–10 calories of energy have been used, mainly in the form of fossil fuels. This goes back to the green revolution that happened around half a century ago. The green revolution saw the yields of many food crops increase tenfold per hectare. This explosion of land productivity was achieved by feeding up to 50 times more external energy into the process of growing food, in addition to breakthroughs in plant breeding. This energy was embedded in fertilizers, pesticides, and such products, as well as mechanized

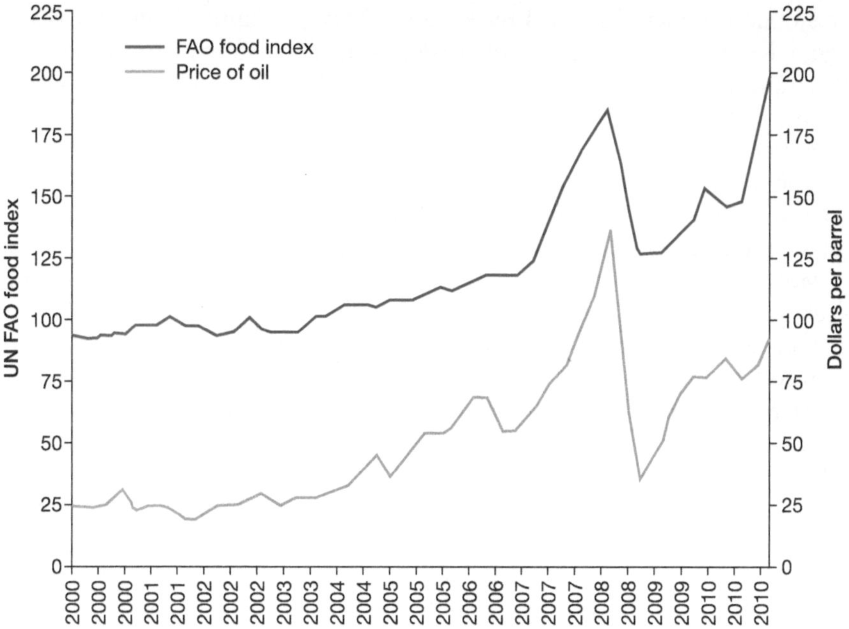

Figure 1.1 Food prices and oil prices, 2000–2010

The prices of oil and food have gone hand in hand during this century.

Sources: Figure adapted from http://richardheinberg.com/235-soaring-oil-and-food-prices-threaten-affordable-food-supply, data from: U.S. Energy Information Administration (EIA) and Food and Agriculture Organization of the United Nations (FAO).

farming, transport, and storage. This all means that when the price of oil rises, so does the price of food.

As well as food, most of our other ways of producing energy are more or less dependent on oil. Coal is mined and transported with machinery that burns diesel. Wind turbines or solar photovoltaic panels use parts that are made from materials that have to be mined, transported, and manufactured using oil. The turbines and panels have to be transported to sites using roads and trucks, and even the factories have been built using machines that burn oil. Bioenergy in its many forms is mainly harvested using machines that run on diesel, and then it gets burned in power stations that were built with oil, and so on.

Oil and the economy

> Economic growth is a function of energy consumption, full stop.
>
> Jeff Rubin, *The End of Growth* (2012)

Historically, our well-being has been closely correlated with our ability to use external energy sources. For millennia the level of population and its economic productivity was pretty stable. This was until about 300 years ago, when mankind started to increase its use of coal. Somewhat ironically, this most likely saved the forests of Europe back then, which were suffering from excessive logging. Coal was the main source of power for the then emerging industrial revolution. Oil has since taken the role of both the biggest and by far the most important energy source in the modern world. Since oil is one way or another embedded in most other goods and products, a rise in oil prices will cause other prices to increase as well.

Most notably this effect can be observed in essential goods and services such as food, energy, and transport. These are necessities that people use in their everyday lives to live and function in society. Since people cannot stop consuming these things, an increase in the price of essentials will create pressure to lower voluntary consumption of non-essential goods and services. This consumption of non-essentials, on the other hand, is one of the basic pillars that modern economies rely on for continued economic growth. As these industries see sales decreasing, they start to postpone investments, lay off people, and cut costs in general. This leads to people having even less money to spend, and the situation worsens even more. For example, in the United States the price of gasoline rose from 75 to 95 cents per liter during the years 2010–2011. It resulted in 280 million more dollars being spent on gasoline every day. This money got cut from consuming other, less essential goods and services.[5]

A substantial amount of the money spent on oil does not end up in the pockets of people that would reinvest it within the country where the money was spent. Instead, these petrodollars generally end up in the hands of a relatively small group of well-off elite. This elite includes the ruling family in Saudi Arabia, Russian oligarchs, dictators, and leaders of developing oil-producing countries, and of course the executives and owners of the international oil companies. Each of these groups will use and invest the petrodollars, but a substantial amount will find its way to arms deals, social programs designed to keep local populations from revolting, subsidizing local gasoline prices, luxury yachts, private jets, and other luxury goods. As the price of oil rises, more and more wealth gets shifted from the many (oil users) to the few (oil producers). In Italy, for example, imports of oil have decreased by 330,000 barrels per day (bpd) since 1990, but the total cost of oil imports has increased from $12 billion to $55 billion. At the same time the trade balance of Italy has weakened by $58 billion. Almost 80 percent of that can be directly explained by the rise in the price of oil.[6] Italy used 1.8 mbpd of oil in 2009.[7]

From the viewpoint of a sole oil-importing country, the situation can be summarized like this: to balance the trade balance, the country needs to increase the value of its own exports by the same amount as the oil price increase. This is a tall order, since all the other oil importers are aiming to do the same and much of the trade happens between oil importers. As the price of oil has multiplied several times in the twenty-first century, so has the speed of transferring wealth from the oil importers to oil exporters. This should be better accounted for when nations and communities calculate the costs of building a local, oil-efficient infrastructure. Often these costs stay within the community instead of being transferred abroad as the petrodollars are. Otherwise, in due course, we will end up selling all that is valuable to us in exchange for oil.

Oil and GDP

The International Monetary Fund (IMF), and Goldman Sachs have estimated that a rise of 50 percent in the price of oil will cut back 1–1.5 percentage points of GDP (gross domestic product) in the oil-importing countries. In the summer of 2012, the IMF forecast that the global economy would continue growing at around 4 percent annually. This would translate to a 3 percent rise in the consumption of oil. As will be shown later in the book, it seems very unlikely that production growth would be able to meet this growth in demand. The remaining options for sustained growth of the world economy are comparable gains in the oil-efficiency of the global economy, and a rapid rise in the development and supply of alternative fuels.[8]

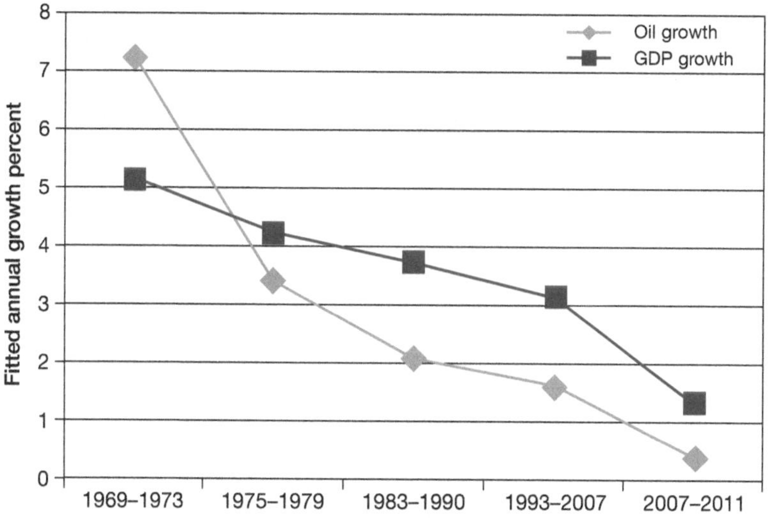

Figure 1.2 World oil supply growth vs world GDP growth

Showing growth in production of world oil and its close correlation with the growth of world GDP despite the economy becoming more oil-efficient (based on exponential trend lines fitted to values for selected groups of years).

Source: Ourfiniteworld.com. Data: World GDP based on USDA Economic Research Data.

As the supply of oil fails to meet demand, the price of oil rises. This leads to demand destruction for some, and higher prices for others. When demand is destroyed, it means that the economic activity that demand would have facilitated is also destroyed. Where is this demand destroyed from, and what activity will be left undone? How rapidly will the effects of this phenomenon spread to the extended economy? Has demand been destroyed more or less justly, or has it been distorted by different subsidies and regulations?

Developing countries use less oil per capita than developed countries. This broadly means that they can generate more economic activity with their "next barrel of oil" than the heavy users in OECD (Organization for Economic Cooperation and Development) countries. This is one of the reasons why many developing countries, such as China, can afford to pay more for their oil than many western nations. The combined oil use in Japan, the European Union, and the United States has been steadily declining for several years now. At the same time their economic growth has more or less stagnated.[9]

According to the U.S. Energy Information Administration (EIA), by 2012 the consumption of oil[10] in thirty OECD countries[11] had declined from its all-time high in 2006 by 6 mbpd. This represents an annual drop of 1 mbpd.[12] The countries that have suffered most in the eurozone (Portugal, Italy, Ireland, Greece, and Spain) also have the biggest shares of oil in their energy mix.[13] Seen from the point of oil, it is no big wonder that their economies are in trouble. Of course part of the decline in oil use is due to gains in efficiency; European cars have ever stricter emissions targets and better fuel economy, and the use of heating oil is dropping in many countries. It would, however, be rather optimistic to assume that the drop in oil use is solely due to more efficient use of oil. The effects of a shattering economy in one country will also be felt in its trade partners, if indirectly.

The most important strategic resource

From a geopolitical or an economic view, oil is currently the most important strategic resource on earth. In the Second World War, it was the availability of oil that dictated the strategies and timings of many actions taken – such as the German assault against the Soviet Union or Japan's attack on Indonesia.[14] Later, the price of oil and therefore the income gained from it played a crucial role in the struggle between the communist and capitalist systems in the 1980s. The crash in oil prices then played a major role in the downfall of the Soviet economy, and society along with it. Now the price increase of the same oil is playing its part in helping Russia to flex its geopolitical muscles once more, growing its influence in world politics.

Oil deposits are distributed on a geological basis, without regard for national borders or the needs of people. The Near East, for example, is said to hold around 60 percent of proven global crude oil reserves. The largest oilfield in the world, Ghawar, is located in Saudi Arabia. Almost all modern humans need oil, even if it is not present inside their national borders. More than 90 percent of the oil used by humanity has been used after 1960. This is a major reason why oil, and the shared global dependency on oil, has been growing in importance in global geopolitics and, with it, grow the risks.

As the thirst for oil in countries such as China continues to grow, the geopolitical situation is likely to get ever tighter. China is currently securing its future oil imports by doing bilateral trade agreements with different producers. Saudi Arabia, for example, has been promised a 49 percent share in new refineries being jointly built in China, effectively securing future supplies of crude oil for the refineries. The leaders of all countries will naturally aim to secure oil for their citizens. As oil production fails to meet demand, this could conceivably lead to tightening international relations and bilateral deals made under the table. Ultimately, the oil will be had by nations and buyers that are most capable, militarily or otherwise, of securing it for themselves.

We conclude with one more final point about oil which is crucial for the message of this book: *Oil is a non-renewable resource.*

In absolute terms, there won't be any more of it, at least, not within a timescale that matters to human civilization. When a barrel of oil is burned, it is burned for good. We have started with the easy resources. As we proceed, oil will get harder and harder to produce. New oil will be located in remote places, difficult and expensive to find and produce. It will be of lesser quality and will have worse net energy for society. Environmental damage could be higher, as will greenhouse gas emissions.

The world has used more than 95 percent of its oil since the beginning of the Second World War. We used over 60 percent after 1960, and over 43 percent after 1990.[15] By 2013 we had burned through around 1,300 Gb (gigabarrels, or billion barrels) of oil. Each year we use more than 30 billion barrels more. Mankind – you and me – should start thinking what we will do as the supply of liquid fuels starts to decline, but demand keeps increasing. One day, oil will not be readily available for an affordable price in the quantities that we need it. That day is most likely nearer than many would like us to believe. The international media has been showing signs of grasping the situation, but we might be years, if not decades, too late. The decline of petroleum is very close.

What is peak oil?

> The term peak oil refers to the maximum rate of the production of oil in any area under consideration, recognizing that it is a finite natural resource, subject to depletion.
>
> Colin Campbell, former oil geologist and the founder of ASPO International

Peak oil is often misconstrued as running out of oil. Sometimes it feels that this muddling is done on purpose – a simple rhetorical trick (a straw man) where one claims that their opponent is saying something that is not true, which can therefore easily be discounted as silly nonsense.
Peak oil is NOT about running out of oil.

> We are not running out of oil, but we are running out of oil that can be produced easily and cheaply.
>
> James Murray and David King (2012), *Nature* 481, 433–435

The moment of peak oil is difficult to pinpoint, as it can only be seen in the rear-view mirror, after production has been declining for a few years. However, if the price of oil has quadrupled in less than a decade with few gains in new production, we can be fairly confident that we are near the peak. Some have said that we will have a prolonged bumpy plateau in oil production as we hit the peak, and currently this seems to be happening.

There have been several peaks around oil already:

- peak in oil discoveries around 1966;
- peak in oil produced per person around 1979;
- peak production of the five large international oil companies in 2004;
- peak availability of net exported oil (less China and India) in 2005;
- production peak of conventional crude oil in 2006–2008.

Forecasting the peaking of oil production is difficult, since a host of uncertainties affect oil production. These include consumer behavior, economic growth, oil prices, technological progress, weather, regulation, local societal crises and events, geopolitics, war, incomplete information of oil markets, and mistakes in geological estimates and measurements, just to name a few.

In addition, there are several variables in defining what peak oil really means that can cause confusion and make discussion harder. These include:

- What is counted as oil? Are we talking about conventional crude pumped onshore? Or do we include almost everything that burns, such as extra heavy oil, bitumen, biofuels, kerogen, liquefied coal, and natural gas liquids?
- What is the time period used? Is it 1 year, 1 month, 1 week, or 1 day of production?
- What is the cost, in terms of energy, money and environmental damage, that is acceptable? Some analysts are complaining that, for example, environmental regulations are to blame for the problems of oil production. They tend to forget that humans have needs other than maximizing the daily production of oil at any price.

- Will reduction in demand also lower the need to produce oil, and would it really be peak demand, not peak production? This is a somewhat naive view, as the reduction in demand is due to higher prices, which are due to the inability of production to meet demand at the lower price.
- What is sufficient accuracy in predicting peak oil? The only certain thing is uncertainty. Therefore it is much more important to understand that the exact moment of peak oil does not matter that much. What matters is that societies should start weaning themselves off oil use, probably decades before actual peak oil happens.

A short history of the theory of peak oil

A geologist, Marion King Hubbert, developed the theory of peak oil in the 1950s. He used it to predict that the peak of conventional crude oil production in the United States would happen between 1965 and 1970. This forecast for the peak of U.S. oil production was widely dismissed and ridiculed, until it proved to be correct: the production of oil peaked at a little over 10 mbpd around 1970. When trying to predict global oil production, Hubbert had used a much too low estimate for total oil reserves, with a far too low figure for total production. Therefore his forecast for the peaking of global oil production has since been proved wrong. In addition, many factors outside geology were affecting oil production and consumption, such as the oil crises in the 1970s.

One of Hubbert's key observations was that in order to produce oil, it must first be found. In other words, it could be said that as the rate of new discoveries started to fall, it was possible to predict when production from those sources would also start to diminish. Hubbert's model is a mathematical model that focuses on conventional oil production. It does not account for external factors affecting the real world such as the price of oil, development of production technology, and the additional reserves and unconventional oil that can be exploited by new technologies. However, Hubbert's model, pictured as a bell curve of rising and falling production, for all its faults, has held its ground decently against many real-world production curves.

There have been many others predicting the date of peak oil since Hubbert's day. The estimates usually range between the years 2000 and 2050. There have even been those who say that peak oil will never happen. There are many problems with the estimates on what is counted as oil production. For example, the so-called refinery gains add to the volume of petroleum products, but it would be a far stretch to call it actual "oil production."

Different agencies and organizations also define oil differently. As noted, the peak in conventional crude oil most likely happened in 2006. But if we count all other liquid fuels, the peak might still be in front of us. Many analysts place the peak of all liquid fuels between 2015 and 2018. This would also mean that the peak volume will not be much higher than it now is, at around 90 mbpd. It is somewhat academic to argue about the exact peak date of conventional or all liquid fuels. This is because preparations and the building of a radically more oil-efficient society will take decades – after we really decide to start it. Furthermore, the production of all liquid fuels has shifted to a new phase of slower growth since around 2005. The total production has hovered between 85 and 90 mbpd for almost a decade. At the same time, the price of oil has more or less quadrupled. In addition, practically all future growth in the consumption of oil will happen outside the OECD countries.[16]

As increases in oil production become more difficult and costly, consumption in the oil-exporting nations continues to grow. The tightening oil market offers a fertile ground for a host of problems to grow upon. These include resource nationalism and resource wars, increased price volatility and geopolitical instability. A peak in available net exports of oil was experienced in 2005, and as China and India, among others, increase their share, the situation will get worse. It is amazing how these events and developments have been ignored in many countries and the media. For all practical purposes, we are living peak oil now.

Peak oil in reports and predictions

> *International Energy Agency chief economist Fatih Birol:* When we look at the oil markets the news is not very bright . . . We think that the crude oil production has already peaked in 2006 . . . The time is running out, the oil is today our lifeline, it is everywhere in the economy, if the prices go up or if there's a supply disruption this will be definitely very bad news
>
> *ABC TV*: How urgent is this?
>
> *Birol*: I think it would have been better if the governments had started to work on it at least 10 years ago.
>
> Fatih Birol in ABC TV interview in 2011[17]

For years, peak oil remained along the margins of media and society. During the last few years, it has crawled its way into more and more official and serious reports. Many of these reports say that the peaking of all liquid fuels will probably happen during this decade – within a few years' time. More specifically, many predict a kind of supply crunch to happen between 2015 and 2018, which will be due more to lack of investments. Given the long lead times of developing oil production those investments needed to be made years ago. This also makes it possible to see the crunch coming in advance. These analyses have not accounted for the rapid rise in shale oil production in the United States, which has quite a short time frame from drilling to market. It will be interesting to see if shale oil mitigates the supply crunch.

It might make the crunch even more sudden, for reasons explained later. In addition, these predictions often assume that everything goes as planned – that Iraq will calm down and manage to increase its oil production dramatically in just a few years, for example.

The International Energy Agency (IEA) was among the first to sound the alarm on the coming supply crunch back in 2007.[18] In 2008 it was followed by Chatham House with their report entitled "The Coming Oil Supply Crunch."[19] The IEA continued its warning in its 2008 energy report, which stated that we would need 30 mbpd of new capacity online by 2015.[20] In 2010 it was joined by ITPOES, the British Industry Taskforce on Peak Oil & Energy Security, which reported that we would lose spare oil production capacity between 2012 and 2015. It was followed by the U.S. Department of Defense's Joint Operations Environment (JOE) report, which predicted that we would lose spare capacity by 2012 and that there would be a 10 mbpd gap between production and demand by 2015.[21] This gap, if it exists, will be made invisible by the rising price of oil which will lower the demand to match production. Demand is always a question of price. In the first quarter of 2012, global spare capacity had dropped by a third from the previous year to 2.4 mbpd. This level, less than 3 percent of total production, was then at its lowest since the price crash of 2008.[22]

And so it goes on.[23] It is a little ironic that many of those critical about peak oil have accused the international media of dwelling way too much on peak oil and its possible consequences. They were offered consolation in 2012, as the IEA published its *World Energy Outlook* (*WEO*) report for the year and the media went crazy. *The New York Times*, for example, told its readers in screaming headlines that the United States would become a net oil exporter by 2030 – a story mirrored by others around the world for months to come. The report itself did not hold such promises. IEA chief economist Fatih Birol stated that the United States would import around 30 percent of its oil in 2035.[24]

Predictions gone wrong

> It is sort of a strange coincidence that the errors seem to be all in one direction.
>
> Gail Tverberg in a discussion about the IEA *World Energy Outlook* report of 2008[25]

It is an often heard (but fallacious) argument that people have been predicting peak oil, or even the end of oil, for at least a hundred years, and failing, so why should we listen to them now when they tell us that the peak is at hand?

First, the fact that some people have been predicting the end of oil in the past has nothing to do with the current situation. This kind of logic would make it unnecessary to slow down at a crossroads if there was no accident the last time. The current situation needs to be assessed by the current information and facts, not by searching for clues from the past.

Second, people presenting this argument often forget that the organizations and reports supporting their own arguments have been categorically wrong in many of their own predictions in the past, and often by a lot wider margin than the modern peak oil predictions. While this will not make their current predictions wrong, it does offer a reason to investigate their methods more closely, for example. Instead of arguing who is right and who is wrong, we should concentrate on finding out what the right answer is.

Recent developments have been quite clear. More and more reputable organizations and reports have increasingly been raising their voices about peak oil. At the same time, the optimists have been dragging their future predictions on oil production to ever lower levels. It seems that we have been on a production plateau for almost a decade, with very little growth in oil production. At the same time, the amount of net exports to the open markets has been in decline. Most OECD economies have also been on a slow growth plateau during the last few years. From that perspective, there is little to win if the only way to lower oil prices is by falling into a recession. At the first sign of growth, the prices of oil start climbing, forcing the economies back down. This will lead to a situation where it will be harder and harder to invest in more oil-efficient infrastructure.

Predicting oil production, on the more optimistic end, there are the big oil companies such as Exxon, BP, and Chevron with the OPEC producers. Governmental agencies such as the EIA are also often said to be quite optimistic, with research and consulting companies close to the oil industry, such as Cambridge Energy Research Associates (CERA), counted among the optimists. The common view of these organizations seems to be that peak oil will happen around or after 2030. What adds an interesting flavor to their predictions is that many of these organizations are directly or indirectly dependent on humanity not worrying too much about oil, or doing anything to end their oil addiction.

Even these optimists have not failed to notice the evolving trend of very slow growth in production capacity. Behind the scenes, the international oil companies have been fighting tooth and nail to replace their oil reserves with new additions. Increasingly they have been doing this by buying smaller oil companies and their reserves. This, of course, does not add to the global proven reserves, but merely shifts them around on paper. The other route has been to add more natural gas reserves, usually noted as oil equivalents. Worryingly, in recent years even the optimists such as the EIA and CERA have revised their estimates on future production growth from the 2-plus percent annually to closer to 1 percent annually.

The Energy Information Administration

The EIA is the statistics and analytics arm of the U.S. Department of Energy. It publishes an Annual Energy Outlook report. Its predictions have historically been on the more optimistic side, but it is interesting to observe how they have changed during recent years. During the last 10 years or so, its estimate for oil production in 2020 has dropped from 120 mbpd to close to 90 mbpd. If this trend continues, its estimates will meet reality at around 50–70 mbpd in the actual year 2020.[26]

The new trend of slow growth in oil production can be seen from the public statistics of the EIA.[27] Crude oil plus condensates have grown less than 0.5 percent annually since 2005. Even this growth is mostly from shale oil and bitumen from the oil sands, and it is perhaps a stretch to call it conventional. Different organizations have different categories and names for different kinds of oil.

In 2013, the EIA published a report on global shale oil and shale gas resources[28] in which they estimated that the global shale oil resources were 345 billion barrels, which is more than 10 years of total consumption. Many undoubtedly got the idea that we had just found a huge amount of oil, especially with headlines saying that there was much more shale oil and shale gas than previously estimated. Fewer people read the conditions and specifications of these resources. They are unproven, which means that nobody has actually found the oil yet, let alone done much drilling to assess the price of production. The amount has been estimated by extrapolating from similar areas currently producing in the United States. Given that an inherent property of shale basins is that they vary immensely in their quality, it becomes even riskier to build a nation's energy future on shale. David Hughes, a widely respected geoscientist and fellow at the Post Carbon Institute, estimates that maybe 5–10 percent of total shale oil resources can be produced at current prices and technology.[29] Ten years of speculative supply quickly shrinks to less than 1 year of actual supply.

Cambridge Energy Research Associates

CERA is a consulting house that is said to align quite closely with oil companies. Among its many analysts is Daniel Yergin, who has published two award-winning books on energy, *The Prize* (2008) and *The Quest*

Table 1.1 Growth in different categories between 2005 and 2012

Product/Million barrels per day	*2005*	*2012*	*% change*	*% annually*
Crude oil + condensate	73.6	75.6	2.59	0.37
All liquid fuels (incl. NGPL + biofuels)	82.3	86.7	5.20	0.75
All liquid fuels + refinery gains	84.4	89.1	2.96	0.77

NGPL = Natural Gas Plant Liquids - Refinery gains mean increases in volume due to the refining process.

Source: EIA International Energy Statistics.

(2011). CERA has been known for its rather optimistic predictions for oil production capacity, but even these predictions have seen an interesting trend developing. CERA also often just talks about production capacity, which is a little different from actual production. It is also somewhat harder to know the exact figures. The daily amount between actual production and production capacity is called spare capacity. This capacity is crucial in the oil markets. The smaller the spare capacity, the more volatile the oil prices are to even small disturbances in the marketplace.

Some predictions given by CERA during recent years:

- In July 2005, Daniel Yergin states in his article for *The Washington Post* that the production capacity of oil will increase to 104 mbpd by 2010. This means a 3 percent growth annually between 2004 and 2010.[30]
- In September 2006, Robert W. Esser, the director of CERA, predicts in *BusinessWeek* that oil and natural gas liquid production capacity will grow by 25 percent by 2015. This equals 2.5 percent growth annually.[31]
- In September 2011, Daniel Yergin predicts in his article for *The Wall Street Journal* that the production capacity for oil and other liquids will grow from the previous year's 92 mbpd to over 110 mbpd by 2030. This equals around 0.9 percent growth annually.

If CERA continues down this path, even their predictions will start to look pretty gloomy in a few years.

British Petroleum (BP)

BP publishes an annual "Statistical Review of World Energy,"[32] which is one of the most comprehensive public reports on reserves and oil production. It is too bad that it relies quite heavily on unconfirmed knowledge in its estimates of reserves.

In the 2013 report, BP states that there are 1668.9 billion barrels of oil reserves. This includes condensates, natural gas liquids (NGL) and crude oil, which also includes heavy oil and bitumen. Here are few observations on those reserves:

- 1187.1 Gb (billion barrels), or around 71 percent of the reserves are located in OPEC countries, of which it is difficult to get reliable, independent estimates.
- Around a quarter of the reserves in OPEC countries (approximately 300 billion barrels) are so-called "political reserves," which are based on one-sided announcements that have never been confirmed. (More on these later in the book.)
- World reserves include 168 Gb of bitumen in Alberta, Canada, and 220 Gb of extra heavy oil in the Orinoco Belt, Venezuela. Both of these are

slow and expensive to produce. These 388 Gb amount to around a quarter of the total reserves.

- In 1992, world reserves amounted to 1039.3 billion barrels, and we have consumed around 600 billion barrels since then.
- In 2002, world reserves amounted to 1321.5 billion barrels, and since then we have consumed more than 300 billion barrels.
- Of the reserve additions since 1992, which amount to around 600 Gb, around 60 percent comes from bitumen and extra heavy oil mentioned above.
- The rest of the additions come from Africa (69 Gb), the Middle East (146 Gb), and Europe and Eurasia (62 Gb, mostly from Russia, Azerbaijan, and Kazakhstan).

A few observations on the development of production, imports, and exports:

- World oil production grew by 1.9 mbpd (2.2 percent) from 2011. Most growth came from Libya (an additional 1 mbpd, which almost reached its pre-Arab Spring level) and North America (the U.S. produced an additional 1 mbpd, and Canada, 0.2 mbpd).
- Total world production was 86.15 mbpd.
- The total output of Canada increased 0.883 mbpd, or 31 percent, between 2002 and 2012. Increasing the production of bitumen from the oil sands seems to be slow.
- The production of bigger European producers has been dropping fast since 2002; Norway by 42.5 percent and the UK by 61.2 percent. Europe imports most (around 10 mbpd) of its 14 mbpd of oil consumption.
- The United States produced around 8.9 mbpd in 2012, an increase of almost 14 percent from the previous year. Still, with consumption at around 18.5 mbpd, the United States imported oil of around 9.5 mbpd.

Consumption has been developing as follows:

- The total European consumption was 14.2 mbpd in 2012, down 1.3 mbpd (8 percent) since 2002.
- Consumption in EU countries has decreased from its maximum in 2006 by more than 10 percent.
- The biggest losers from their all-time highs in Europe are the Republic of Ireland (down 33.3 percent), Portugal (down 31.9 percent), Italy (down 31.6 percent), Greece (down 29.2 percent), and Bulgaria (down 26.7 percent).
- From its all-time high in 2005, the United States had decreased its consumption by over 10 percent by 2012.

Oil geologist Jean Laherrère criticizes the previous BP report (from 2012) in his article on the website "The Oil Drum."[33] Among his concerns are the reporting of backdated reserve additions (more on that later) and the inclusion of bitumen in the oil sands under the category World Oil, when it used to be under its own category. According to Laherrère, if the problem is the size of the tap, the size of the tank does not matter much. He also criticizes the use of unconfirmed, one-sided announcements for oil reserves. For example, Iran and Iraq have been competing on their proven (1P) reserves for years. So much so, that both currently have more 1P reserves than they have technical proven and probable (2P) reserves. It stretches the imagination, but this means that they have more proven reserves than they have proven and probable reserves put together. Stretching it a little further, BP seems to swallow these announcements without question.

The International Monetary Fund (IMF)

The IMF published an interesting working paper called "The Future of Oil: Geology versus Technology"[34] in the summer of 2012. In it, the authors presented their new model on the production of oil. According to the report, traditional economic models, mainly based on the demand of oil (such as from the IEA, EIA, and so forth), had been overly optimistic. On the other hand, Hubbert's model, based solely on geology, had ignored the effects of technology and prices, which led it to be too pessimistic. The new model by the IMF attempts to account for both the geological restraints and the development of production technology and higher oil prices. When compared against historical data, the model has given the most accurate results on both the rate of production and the level of prices. The report is a most interesting, if unofficial,[35] initiative from the IMF.

However, the predictions the model gives are not very encouraging. It forecasts a growth of oil production of less than 1 percent annually for the next 10 years. This, however, requires that the prices of oil nearly double from 2012 levels. Historically this seems unlikely, since the world economy will probably go into recession before we reach those high oil prices. Recent statistics and events even suggest that the current prices of around 100–120 US$ per barrel are enough to tank OECD economies and give even China and India serious problems.[36]

The model also tries to account for non-linear effects. At a certain price point, a 10 percent rise will endanger vulnerable industries such as aviation and tourism. If certain industries crash, they will cause much bigger, and harder to predict, effects on the whole economy than would be apparent just by looking at the price rise itself. These price points are the practical, short-term, maximum prices for oil.

Later the same year IMF published another working paper titled "Oil and the World Economy: Some Possible Futures."[37] It presents some scenarios with different assumptions on some important variables. These include the amount of growth in oil production and its importance and irreplaceability in the economy. The scenarios with low growth in production but high irreplaceability lead to price rises of more than 1,000 percent. This seems impossible. The rises in prices are likely to be halted by something way before they reach such levels. This "something" probably won't be anything pleasing, positive or predictable.

In 2013, the IMF started to work with Uppsala University's Global Energy Systems (headed by Kjell Aleklett of ASPO International) to do further research on how to include peak oil and its effects on economic models.[38]

Organization for Economic Cooperation and Development

The economic department of OECD published a working paper "The Price of Oil – What if it starts rising again?"[39] in the spring of 2013. The model used in the paper suggests a substantial rise in the price of oil by 2020 if the global economy recovers even close to the growth it had at the beginning of the century. Since the variables are many and hard to predict, the price range given is also quite wide – from 150 to 270 US$ per barrel. The baseline scenario means a price rise to 190 US$. The model is economical, and it does not account much for geological restraints on production, but concentrates instead on the demand for oil.

What is worrying is the assumption of growing world oil demand even if prices do rise to 190 US$ per barrel. The model assumes that production will grow by 14 mbpd by 2020. At the same time, the current oil wells will be depleted by at least 12 mbpd. This means that we will have to bring a total of 26 mbpd of new production online by 2020. This is much faster growth than we have experienced in the last 8 years on average. If this growth in production does not happen, the model predicts even higher prices for oil. If growth is 2 mbpd less, the price will rise by a further 15 US$.

Furthermore, if the production of oil indeed rises as the report says, the bulk of the rise will be consumed in developing countries, leaving OECD countries with dwindling supplies. The growth of developing economies is much more oil-intensive, compared with OECD countries. One more interesting point in the report is the inelasticity of the oil market. In the short term, most countries will not see any reduction in their oil use even if the price rises – this means the demand is inelastic. In the mid to long term this elasticity is around 0.2, which means that a 1 percent rise in oil prices will reduce demand by 0.2 percent. With the fourfold increase in price and little growth in production to show for it, it could be said that the production of oil is also quite inelastic. The invisible hand of the market seems to be grasping at nothing, since there is no viable alternative to oil.

The International Energy Agency

The IEA is the watchdog of energy security in OECD countries. It is also the supreme authority for many officials and national agencies of its member

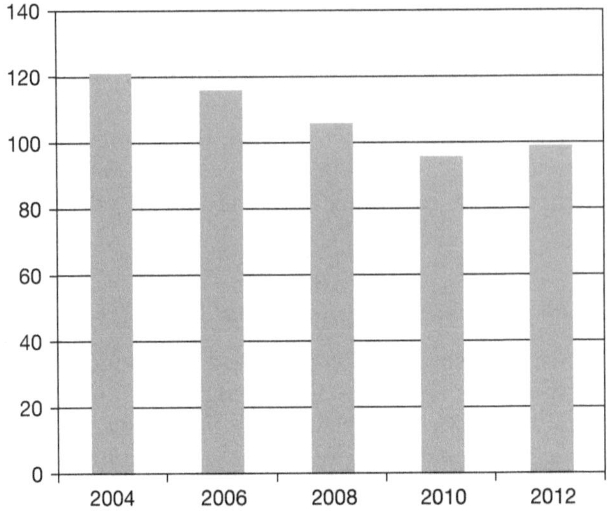

Figure 1.3 The development of IEA predictions for world oil demand in 2030 or 2035 (mbpd)

The IEA has been revising its prediction for 2030 or 2035 total oil demand worryingly downwards since 2004, mainly due to the realization of the fast depletion rates of current oilfields.

Source: IEA.

countries. Its reports and predictions end up in the analyses and plans of officials and analysts, who see them as a safe bet. They can't be wrong, or at least they can't be accused of being wrong, if the IEA said it was so.

Even though the IEA is predicting a slow growth in oil production until 2035, it has been steadily adding conditions that need to be met for the growth to materialize such as massive investments in production and energy efficiency, overhauling systemic corruption in Russia, stable societies in oil-producing nations, stable and convenient oil prices, and so on. It can clearly be seen that we are moving towards the limits of oil production. If all these conditions are met, production would rise to around 100 mbpd in 2035 (WEO 2011 and WEO 2012). In the process of writing this book, the authors got the chance[40] to comment on WEO 2012 and possible problems in oil production that could be addressed in the report. The report, or the media coverage it got, concentrated on the rising shale oil and shale gas production of the United States. Many have criticized the 2012 report as way too optimistic.[41] [42]

In 2010 the IEA took a big step when it moved the peaking of conventional crude oil from the distant future (the 2030s) to the past, to the year 2006. A central mandate of the IEA is to watch the production of oil and its development in the world. Even though it might be better late than never, the IEA's ability to perform as the oil watchdog could be questioned by this lateness. At the same time, the carpet under humanity's prospects and future plans was crudely removed, since that future was planned on the continued and growing availability of cheap oil.

The UK Industry Taskforce on Peak Oil and Energy Security (ITPOES)

ITPOES published its second report on peak oil in 2010. Behind this group and report are companies and individuals such as the billionaire Richard Branson (Virgin Airlines) and Ian Marchant (CEO of Scottish & Southern Energy). The group is worried by the lack of peak oil awareness of the British government. In 2011 the British Department of Energy and Climate Change announced that they would develop a response plan in cooperation with ITPOES for the coming oil shock and the price rises it would bring.[43]

Their 2010 report offers a sobering outlook. It says for example:

- The current growth in production will start to wane between 2011 and 2013, and start to decline soon after. Since it is slow to bring new production online, it is hard to avoid this in the next few years.
- Oil producers are currently not finding enough new, large oilfields.
- The proven reserves of OPEC countries remain unproven by third parties and are therefore unreliable. This calls into question the overall reliability of current predictions.
- The UK government faces two challenges. First, it needs to recognize the problem. Second, it needs to implement policies and response plans in order to avoid the worst consequences. If this is not done, the next government will face a growing risk of endangered energy security and possible shortages of consumer goods. These will lead to problems in the commercial and industrial sectors and it will deepen the poverty of the already poor.
- A growing portion of the world oil production is expensive. This means that recessions that reduce the price of oil also remove production capacity, starting with the most expensive.

It will be interesting to see how the boom in shale oil will effect these predictions, since the report was largely prepared before the hype around shale started to go global. The report sees peak oil happening around 2015, with production around 92 mbpd. In 2013, production was around 90 mbpd. It also predicts that potential demand could be 10 mbpd higher than production by 2015. This will lead to demand destruction by rising prices in many OECD countries, and eventually the destruction will also reach developing countries. So far China and India have been able to buy all the oil that European countries could not afford to buy due to their economic problems and high oil prices. Oil exporters, on the other hand, are increasing their oil use as prices rise.

The Bureau of Infrastructure, Transport and Regional Economics (BITRE)

The Australian Bureau of Infrastructure, Transport and Regional Economics 117 report commissioned by the Australian government was buried at first, until it was leaked in late 2011. It is a question in its own right as to why it was buried in the first place. One theory is that it would have buried the government's greenhouse gas reduction targets, in case Australians started synthesizing liquid fuels from their coal deposits. Another suggestion is that it was not published because of its inaccuracies, but recent production history shows that its accuracy has not been that far off on the whole.[44] It is also worth noting that bringing this type of subject to the public discourse is not very pleasant for most of us, especially politicians. The report uses modified Hubbert linearization and studies production statistics from the world's oilfields quite thoroughly. It ends up predicting peak oil for 2017 and a quite fast decline thereafter.[45] A special worry in the report is the production capacity of Russia, which would lose almost a third of its production capacity in this decade. This is in contradiction to Aleklett's analysis (discussed later), which states that Russian production will stay pretty stable for this decade, assuming enough investments are made.

UGES and ASPO

Uppsala Global Energy Systems (UGES) is a research group at Uppsala University in Sweden. It operates under Professor Kjell Aleklett, who is also the president of ASPO International (the Association for the Study of Peak Oil and Gas). The group has published over twenty scientific articles on the depletion of oil and other fossil fuels. Kjell Aleklett published a book, *Peeking at Peak Oil* in 2012, in which he brings together the findings of their studies.

Aleklett and Campbell (another ASPO founder) published their article "The Peak and Decline of World Oil and Gas Production" in the science journal *Minerals and Energy* in 2003.[46] In the article the authors place the peak at around 2010 with a production of 85 mbpd. It was the first peer-reviewed article about the peaking of world oil production. Compared with actual production history, the prediction has been quite realistic. So far the peak in oil production has been in 2013 with a production of around 85.3 mbpd, according to EIA statistics.[47] So we reached the predicted peak three years later than predicted. It will be interesting to see where we go from here.

During the writing of this book, the most recent production information from the EIA for the 2013 average was:

- crude oil plus lease condensates: 76,053,548 bpd;
- plus all natural gas plant liquids (NGPL): 85,352,546 bpd;

- plus other liquids (such as biofuels) and refinery gains: 90,332,721 bpd.[48]

Bundeswehr and Joint Forces Command

The analytics department of the German *Bundeswehr* published its now famous report on the effects of peak oil in 2010. It has since been translated into English and is freely available on the internet.[49] The report paints quite a worrying picture on the risks associated with peak oil. Many of the changes, such as geopolitical shifts in power from the oil buyers to the oil producers are not distant risks, but are already a progressing reality. The increasing likelihood of civil unrest and resource wars create an unstable and dangerous basis for global politics. The report does not give a date on the peaking of oil production. It mainly explores its many risks and effects on societies in the event that the peak occurs in the near future and catches humanity largely unawares.

The Joint Forces Command in the United States, which was discontinued in 2011, cast oil and energy in a major role in their 2010 Joint Operations Environment (JOE) report.[50] Even though the report predicts that coal and oil will continue their dominance as energy sources into 2030, it also warns about the risks of oil production. It predicted a non-existent spare capacity for 2012 and a 10 mbpd gap between production and demand for 2015. The main risks seen are the lack of drilling rigs, a lack of skilled personnel and insufficient refining capacity. As this book is being written, we know that the first prediction has not come through. Growing production of shale oil in the United States has seen to it that there is more spare oil production capacity in the world than there has been for years. This could change fast, as crises such as the one in Syria can proliferate and disturb oil production and transportation. Also the gap between oil production and demand will probably be smaller than predicted. Given the persistently high oil prices in mid-2013, the gap is still there, only smaller.

Notes

1 http://en.wikipedia.org/wiki/Petroleum_industry.
2 There is an alternative theory for the birth of mineral oil. This is called *abiotic theory*, according to which oil is formed by geological mechanisms deep underground all the time. Both the conventional theory and the abiotic theory can be correct. It is however extremely unlikely that oil is formed at anything close to the rate mankind is using it, biotic or abiotic. More on this matter can be read in the article: Höök, M., Bardi, U., Feng, L., Pang, X. (2010), Development of oil formation theories and their importance for peak oil, *Marine and Petroleum Geology*, 27, 9, 1995–2004, URL: http://dx.doi.org/10.1016/j.marpetgeo.2010.06.005.
3 Nitrogen-fertilizers are made by combining the hydrogen atoms taken from methane with nitrogen under pressure and heat.
4 Downey, M. (2009), *Oil 101*, Wooden Table Press LLC.

5 Murray, J., King, D. (2012), Oil's tipping point has passed, *Nature*, 481, 433–435, doi:10.1038/481433a.
6 Murray, J., King, D. (2012), *Nature*, 481, 433–435.
7 www.indexmundi.com/g/g.aspx?c=it&v=93.
8 Murray, J., King, D. (2012), *Nature*, 481, 433–435.
9 http://ourfiniteworld.com/2012/07/05/the-growing-part-of-the-world-in-charts/.
10 Domestic production + net imports of crude oil and condensates.
11 Excluding Israel, Slovenia, Estonia, and Chile, whose info the EIA did not publish for some reason.
12 www.theoildrum.com/node/9314#comment-903939.
13 www.eurotrib.com/story/2010/4/8/55346/04864.
14 http://en.wikipedia.org/wiki/Netherlands_East_Indies_campaign.
15 Gilbers, R., Perl, A. (2010), *Transport Revolutions*, New Society Publishers, p. 111.
16 IEA (2011), *Medium-Term Oil & Gas Markets*, International Energy Agency.
17 www.youtube.com/watch?v=iKkISqOCnVA.
18 www.davidstrahan.com/blog/?p=73.
19 www.chathamhouse.org/sites/default/files/public/Research/Energy,%20 Environment%20and%20Development/0808oilcrunch.pdf.
20 www.iea.org/weo/docs/weo2008/WEO2008_es_English.pdf.
21 www.jfcom.mil/newslink/storyarchive/2010/JOE_2010_o.pdf, p. 29.
22 www.eia.gov/todayinenergy/detail.cfm?id=6410.
23 www.energybulletin.net/node/52460.
24 http://aleklett.wordpress.com/2012/11/29/an-analysis-of-world-energy-outlook-2012-as-preparation-for-an-interview-with-science.
25 http://europe.theoildrum.com/node/4829#comment-440932.
26 www.theoildrum.com/node/9182#comment-893290.
27 EIA International Energy Statistics – www.eia.gov/cfapps/ipdbproject/IED Index3.cfm?tid=5&pid=53&aid=1.
28 Technically recoverable shale oil and shale gas resources: an assessment of 137 shale formations in 41 countries outside the United States, www.eia.gov/analysis/studies/worldshalegas.
29 www.guardian.co.uk/environment/earth-insight/2013/jun/21/shale-gas-peak-oil-economic-crisis.
30 www.washingtonpost.com/wp-dyn/content/article/2005/07/29/AR200507290 1672.html.
31 www.businessweek.com/stories/2006-09-17/commentary-plenty-of-oil-ust-drill-deeper.
32 www.bp.com/statisticalreview.
33 www.theoildrum.com/node/9389.
34 Benes, J., Chauvet, M., Kamenik, O., Kumhof, M., Laxton, D., Mursula, S., Selody, J. (2012), The future of oil: geology versus technology, IMF Working paper, www.imf.org/external/pubs/ft/wp/2012/wp12109.pdf.
35 Working papers do not represent the official view of the IMF.
36 http://ourfiniteworld.com/2013/06/07/high-oil-prices-are-starting-to-affect-china-and-india.
37 Kumhof, M., Muir, D. (2012), Oil and the world economy: some possible futures, IMF Working Paper.
38 www.uu.se/en/news/news-document/?id=2631&area=2,5,16&typ=artikel&na=&lang=en.
39 Fournier, J., Koske, I., Wanner, I., Zipperer, V. (2013), The price of oil – will it start rising again?, OECD Economics Department Working Papers, No. 1031, OECD Publishing. http://dx.doi.org/10.1787/5k49q186vxnp-en.

40 Finland is a member of the IEA, and so gets to comment on the report before it is published.
41 www.energyvanguard.com/blog-building-science-HERS-BPI/bid/56391/Why-the-IEA-s-2012-World-Energy-Outlook-Could-Be-Wrong.
42 www.peakoil.net/headline-news/an-analysis-of-world-energy-outlook-2012-as-preparation-for-an-interview-with-science.
43 www.businessgreen.com/bg/news/2072738/exclusive-government-develop-oil-shock-response-plan.
44 http://crudeoilpeak.info/report-card-2012-it-was-a-very-good-year-for-peak-oil-denial.
45 www.manicore.com/fichiers/Australian_Govt_Oil_supply_trends.pdf.
46 www.peakoil.net/files/OilpeakMineralsEnergy.pdf.
47 We are counting crude oil, lease condensates, and natural gas plant liquids as oil. Other liquids, mainly biofuels, and refinery gains are not counted as produced oil.
48 www.eia.gov/cfapps/ipdbproject/IEDIndex3.cfm?tid=50&pid=53&aid=1.
49 For example here: www.permaculture.org.au/files/Peak%20Oil_Study%20EN.pdf.
50 www.jfcom.mil/newslink/storyarchive/2010/JOE_2010_o.pdf.

2 The end of cheap oil

Peak oil dynamics

Many have guessed, predicted, and hoped that, after peak oil, there would be a few decades of a stable, undulating plateau of oil production. This would offer the necessary decades needed to wean us off oil. According to recent years of oil production, we are on that undulating plateau now. According to the IEA *World Energy Outlook 2012*, the growth of oil production in the next decades will be slow. What many fail to notice is the growing list of assumptions including incredibly good luck, stable societal conditions worldwide, a healthy economic situation, and ever-growing investments in oil production that even this slow growth will demand. If even some of these expectations and conditions fail, the whole house of cards could fall down. This setting has been called peak oil dynamics.

These dynamics include a lot of factors that are outside the conventional, geological peak oil theory. Many of these factors are negative from the perspective of oil production, promising growing difficulties in the future. *The Black Swan*, a bestselling book by Nicolas N. Taleb, explains many of these potentially devastating risks that grow on tightening oil markets and the unstable situation that they create. In a globalized world, these risks can cause global havoc and enormous cumulative destruction in a way previously unseen.

A "black swan" can be defined as having three features:

1 it is a surprise for the observer;
2 it causes wide ranging and serious chain reactions, and;
3 in hindsight, it can be rationalized and viewed as predictable.

The following chapter presents many of the components that make up peak oil dynamics, and which make us prone to black swans. In reading this book, many of these black swans will become grey for the reader, as they will no longer be surprises, and one can prepare for them to some extent. The concept of black swans does not rule out the chance of positive surprises but it would be foolish to count on these when planning and building for the future.

EROEI

Energy returned on energy invested (EROEI) is the relationship between the energy invested in energy production, and the energy gained from it. For example, if a barrel of oil needs to be spent to produce four barrels of oil, the EROEI of the process is 4:1. EROEI, also sometimes called EROI, is similar to the concept of return on investment (ROI), a common concept in the investment world. EROEI is one of the most important properties of energy production in the long run. Investments are usually made on projects that have the highest ROI (on similar level of risk). Similarly, societies usually (should) invest their energy in projects that have the best return on that energy.

If our average EROEI is 5:1, it means that 20 percent of our energy goes into producing energy. The remaining 80 percent can be used for other purposes such as food production and processing, education, building infrastructure, manufacturing consumer goods, health care and financial services, and so on. An EROEI of 1:1 on the other hand means that all our energy is used to get that energy with nothing left for other purposes. It would mean that everybody would be working directly on energy production. The closest we have been to this was perhaps some time in our distant history, when most of the people were directly working in hunting and gathering food and a group had only a few, if any, specialists living on other people's surplus.

When EROEI approaches 1:1, the notion of civilization, human dignity, and solidarity disappears. Everybody is looking, indeed has to look, only to his or her own energy requirements and how to fulfill them. You could imagine yourself walking an endless hallway at some imaginary store. Here and there is a piece of bread and some water to be found, but only enough that you can get to the next piece of bread. If you stop for a chat with the fellow walking beside you, or start to make plans about how to make the situation better, you will starve before you reach the next meal. Nothing can be done, because doing anything will mean that your personal EROEI will go below 1:1 and you will die.

Despite its importance, the concept of EROEI has been all but neglected in modern societies. This is probably due to the (temporary) energy glut offered by fossil fuels. EROEI has simply not been that important, as it has generally been large enough for the main energy sources. EROEI will be revisited many times in this book, because the EROEI of modern oil production, as well as other fossil fuels, has been declining steadily.

The minimum average EROEI for a modern industrial society has been estimated to be between 5:1 and 10:1. The production of liquid fuels for the transport sector needs an EROEI of at least 3:1 to 5:1, and this doesn't leave any for the rest of society, for instance education or health care.[1] Quite likely this average EROEI is closer to 10:1 than 5:1 with anything resembling our current economic system. It can even be argued that in addition to needing more and more growth in usable net energy, economic growth also needs better and better EROEI to make it more efficient for economic activity such as producing products and services. Gail Tverberg, a well-known peak oil analyst, speculates that even our current EROEI might be too low for our current economic system. A lower average EROEI such as 10:1 might run a more efficient and differently structured economy, but not the one we have right now, which is also the one we are stuck with for the time being.[2]

We can also view EROEI with each layer of society's needs or complexity adding further demands on higher EROEI. According to a recent report[3] these layers with their corresponding EROEI requirements relate to the following:

- 1.2:1 Oil production and refining
- 3:1 Transportation
- 5:1 Food
- 7–8:1 Support family of workers
- 9–10:1 Education
- 12:1 Health care
- 14:1 Arts and other.

Calculating EROEI is somewhat tricky. There is no widely accepted standard on how to do it and what to include. It also greatly depends on the energy source. Calculations can be made endlessly complicated by including more and more distant and indirect elements. For example, we can apply EROEI to a power plant. In addition to the plant and its infrastructure, should we include the raw materials and their production that was used to make the plant? What about the machinery to mine and process that material? Do we somehow include the factory workers, designers, and executive officers with their education, housing, health care, and food? What about the infrastructure used to make those? And so on, ad infinitum. It is good to recognize that all EROEI calculations are approximations and greatly depend on the variables and assumptions chosen. This is not a big problem as long as the EROEI stays well above 10:1, and the ones using the EROEI know how it has been calculated. This is not often the case. When EROEI drops below 8:1, the net energy for society starts to diminish more and more rapidly. An EROEI of 100:1 means there is 99 percent net energy, 10:1 means there is 90 percent net energy, but 2:1 means there is only 50 percent net energy.

EROEI of oil

Humankind, indeed life in general, tends to go for the low hanging and juiciest fruits first. This is highly rewarding, at least in the short term. Energy production is no exception. The EROEI of producing oil in the early twentieth century has been estimated to be around 100:1. It was highly profitable from a net energy perspective. To get 100 barrels of oil, only

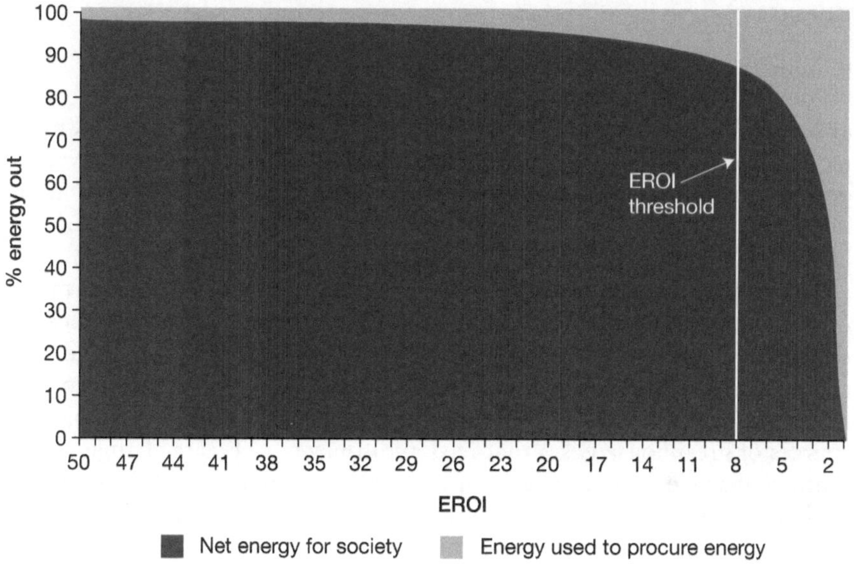

Figure 2.1 Net energy for society drops rapidly when EROEI goes under 8:1

Source: Attribution – David Murphy, www.theoildrum.com/node/8625. This file is licensed under the Creative Commons Attribution-ShareAlike 3.0 United States license.

1 barrel of oil (or its energy content) needed to be used. Nowadays the EROEI of oil production is much smaller, and depends greatly on the location and method of production.

Estimated EROEI of different forms of oil production:

- 100:1 Oil in the 1930s in the United States
- 25:1 Oil in the 1970s in the United States
- 15:1 Oil in the 2000s in the United States
- 4–7:1 Deepwater and ultra-deepwater oil[4]
- 3–6:1 Bitumen from oil sands in Canada
- 2:1 Oil shale (kerogen) production.[5]

So even though we still have around half of our crude oil reserves left, it is the worse half from the perspective of EROEI. It is deeper underground, below deep sea, in the Arctic region, of heavier and poorer quality, in politically unstable regions and so forth. Humanity, or more precisely, our oil companies, must spend more money and resources for exploration, production and transportation of oil nowadays than they did in the 1970s when we discovered the last truly big oilfields. All of this translates to more and more resources and energy being spent, which means that less will be available for the rest of us. It is highly unlikely that the poor EROEI, around 5:1, of most unconventional oil production is enough to run modern industrialized societies. This development alone should ring our alarm bells.

EROEI of coal

In addition to oil, our other main energy source, coal, has seen its EROEI go down. The widest seams in the best places have been produced first, as has the highest quality coal. Every day we have to go deeper and mine more to get coal of lesser quality than before. In addition, the new coal deposits are located in more distant places that require new transportation infrastructure. Measured by energy content, the coal production of the United States peaked in 1998.[6]

The relative abundance and cheap price of coal does not include many of the negative effects and costs of its production and use. These include the hundreds of thousands of deaths annually caused by mining and burning coal, and the wide distribution of toxics such as mercury into the environment and eventually the food chain. It is also one of the biggest sources of greenhouse gases that feed anthropogenic climate change. Some estimate that we have 250 years of coal left on current usage. A lot of this coal, however, will not be economically or technologically feasible for production with current prices, despite the reasonably high EROEI coal has today. The price of coal rose by hundreds of percent in the 2000s. It is often burned quite near to the mine, and so plays a far smaller role in the global energy markets than oil.

The latest, quite ironic, twist in our story of coal is the start of coal imports from the U.S. to Europe. Coal use has been declining in the U.S. as a feedstock for generating electricity, mainly because of cheap natural gas due to a recent gas glut on the market. In Europe, however, natural gas remains expensive. So the leftover coal from the U.S. has been shipped to Europe to burn at local power plants. This has created a somewhat perverse situation, as Europe has been trying to profile itself as the primary champion against climate change.

EROEI of natural gas

The EROEI of natural gas production depends on many things. These are, for example, the method of production, its location, the location of the end use, and the infrastructure between them. If a pipeline is built, moving the gas to market is easy and efficient, but the pipeline is more expensive to build. If there is no pipeline, the gas perhaps needs to be liquefied for transport by trucks or ships. This is made by cooling it to –160°C, which takes more energy than transporting it with a pipeline.

In many locations, natural gas is a by-product of oil. If transportation cannot be economically built, the gas is flared at the site. This happens for example in the Bakken shale oil formation.[7] It can also be pumped back into the reservoir to help pressurize it, as is done in Alaskan North Slope oilfield.[8] Some of the gas is also burned for local needs at the oilfield. The production of shale gas requires constant drilling to offset the rapid decline of shale gas wells. Since the wells deplete fast, there might not be economic sense in building a permanent transportation infrastructure. All these factors make the EROEI of shale gas potentially weaker than that of conventional gas. The picture may be more complex than that, as we see below.

The EROEI of natural gas is estimated to be between 20:1 and 40:1. According to a recent preliminary study, the EROEI of shale gas from the Marcellus basin is much better, in the range of 64:1 to 112:1 with a mean of 85:1,[9] calculated as gas at the wellhead. There are a lot of variables to consider in the study. It uses production data from a quite short period of time, 2.3 years, and assumes that the wells will stay in production for 30 years. This might be unlikely since the production flow could dwindle so low that it would not be worthwhile to keep the wells in production that long. Perhaps most importantly, the best spots of Marcellus shale have largely been drilled first, with these results representing their production. It is highly unlikely that all the estimated total resources could be produced this efficiently. This high EROEI for Marcellus shale gas might represent the easiest fruits that we picked up with oil production around 100 years ago, when the EROEI for oil was 100:1 or even more.

EROEI for other energy sources

Below are a few estimated EROEIs for different fuels and electricity production, which also takes into account the efficiency of turning primary energy (e.g. wood, coal) into secondary energy (electricity). The research into EROEI has gained momentum in recent years, perhaps because we have noticed that our average EROEI is dropping. In estimating electricity production, the supporting infrastructure, such as the need for load following and energy storage (pumped hydro), are also taken into consideration. It is mainly based on a recent peer-reviewed study by Weißbach et al. (2013).[10] This study also offers criticism and points out some flaws in other studies done on EROEI in recent years. This helps to understand the wide range of different EROEI estimates for different energy sources. The main criticisms concern:

- weighing the operational lifetime of power plants – many studies assume absurdly low operational lifetimes for nuclear and fossil fuel power plants and unreasonably long ones for renewable power plants;
- the multiplying of direct electricity production by three so it would compare with the primary energy use of regular power plants that turn heat into electricity (usually with efficiency of around 30–50 percent). This means that EMROI (energy money return on investment)[11] is calculated and then compared with the EROEI of fossil or nuclear fueled power plants. These are two slightly different things;
- not including the need for energy storage and load following;
- the use of old data and methodologies.

The study also notes the intermittency of many renewable energy sources. As such, they need a supporting infrastructure of energy storage and load following power plants (called buffering). This buffering is needed to capture the extra energy when production is high but demand is low and to help during times of low production but high demand.

Fuels (with some ranges in estimates)

- firewood 30:1
- conventional crude oil in the 1990s 15:1
- bitumen from oil sands 3:1 (1.5–5)
- liquid biofuels (corn ethanol in the U.S.) 1:1 (0.5–2)
- liquid biofuels (ethanol from sugar cane in Brazil) 8:1.

Electricity production (without buffering/with buffering)

- solar photovoltaics in Germany 3.9:1/**1.6:1**
- biomass (corn) 3.5:1/**3.5:1**
- wind power (E-66 turbine, 1.5MW) 16:1/**3.9:1**
- concentrated solar power (CSP, in desert) 19:1/**9:1**
- advanced combined cycle gas turbine (CCGT) 28:1/**28:1**
- coal (without transport) 30:1/**30:1**
- hydropower (medium size) 49:1/**35:1** (11–260[12])
- nuclear (current, pressurized water reactor (PWR)[13]) 75:1/**75:1**.

Out of interest, we can look at the case of nuclear power a little closer. There are estimates ranging from 1:1 to 180:1 reported for nuclear. Why? The main reasons are the differences in how to measure the energy used in its construction and production. Many studies show some kind of bias when choosing data and assumptions which distort the results, usually to the detriment of nuclear power. For example, it is unlikely that a single uranium mine could consume more energy than the whole nation in which it is located (yet this is suggested by some). Further, only a small proportion of nuclear fuel today is enriched with the most energy-hungry process of diffusion. The current trend, well on its way, is a move to centrifuges that are maybe 50 times more efficient than diffusion. Also, the operational lifetimes of nuclear power plants are often assumed to be much lower than they actually are. These assumptions might result in an EROEI of well below 5:1 being given for nuclear, but they do not represent the real world in any sense.[14]

Fourth generation nuclear reactors, such as the IFR (Integral Fast Reactor[15]), has an EROEI of perhaps 10 times better than current reactors. This is mainly due to its ability to use the fuel almost 100 times more efficiently. It can use the spent fuel from the current nuclear fleet as its fuel, as well as weapons-grade plutonium or natural uranium (U-238) that has not been enriched. If current spent fuel is burned in an IFR, the waste will be dangerously radioactive for only about 200–300 years, and will also be of lesser volume. This can be compared to the current requirements of maybe hundreds of thousands of years of safe final storage for used nuclear fuel. From an EROEI perspective, this is all great news in a world that has been suffering from worsening EROEI for the last few decades. The impact these kinds of reactors can have on the oil crisis we are facing is quite limited, however, as the technology is not yet widely available and will take decades to build on a large scale.

Energy is the most important thing

Energy is the mother, father, and grandparents of all other resources. Without useful energy, everything else will be left undone. Peak oil with a host of other factors, which will be discussed later, can lead us to a liquid fuels crisis in the short term. In the longer term, we will face a worsening energy crisis due to worsening EROEI for most of our energy sources. Some, like Tim Morgan in his book *Life After* Growth,[16] argue that declining EROEI is one of the main reasons behind current economic problems, and that the situation will get far worse faster than most expect.

Until recently, humanity has been moving from worse fuels to better ones. We started moving from wood to coal in the sixteenth century. Then we started using oil and natural gas in the nineteenth century, and then added uranium to our energy mix in the twentieth century. It needs to be remembered that we did not stop burning wood. We just added more energy sources on top of it. An important feature of the energy production industry is its slowness. The transition to new energy sources always takes decades, if not centuries. It is also important to understand that for the first time in modern history we are largely trying to move from better (conventional crude oil) to worse (substitutes such as biofuels or bitumen) fuels and energy sources. We are also trying to do it in only a few decades, while global energy demand keeps rising. It seems unlikely that we will succeed in this without major problems along the way.

History has not seen many cases when a civilization of significant size has managed to simplify its structure voluntarily, and therefore reduced its energy demand. Byzantium is one of these examples. It managed to systematically reduce the bureaucracy and hierarchies of its society when it was faced with diminishing energy production and imports (mainly grain). When Western Rome finally collapsed after centuries of struggle in the fifth century, Eastern Rome (Byzantium) managed to crawl back from the brink of collapse to a new golden age, and managed to reclaim formerly lost territories. The difference today is that our problems are largely global.

EROEI is one of the most important factors that will make the effects of peak oil worse in the longer term. Even if we manage to produce 100 mbpd of oil in the future, it will be of no great use if we need to put half of that back into the production. Meanwhile, we get all the ill effects of the larger production with accidents and pollution, which also demand energy to be remediated. A diminishing EROEI in a growing global population will leave much less energy for each of us.

The significance of location and quality

We are not running out of oil. We are running out of oil that is close to markets, cheap to produce, and of good quality. All of these have a relationship with EROEI as already discussed. If we have to go to the Antarctic to get oil, we will get less net energy from it. If the oil is of poor quality and more toxic, it will need extensive refining, and we will get less net energy from it.

In the near future, the quality and location of new oil might become a more acute problem than net energy. It has been estimated that there is a lot of oil and natural gas in the Arctic region. Since it is located under ice and snow, it will be harder to explore and produce. Since there has not yet been much in the way of preliminary drilling, these estimates of oil and gas remain as speculation. In a practical sense, we think we have found oil and gas in the Arctic, but so far have failed to prove our finding by actually drilling for that oil. Furthermore, even if the difficulties of production are solved, the oil is only valuable after it has been delivered to the market.

There is a logistical nightmare waiting for oil producers in the Arctic region. Damage to roads due to ground frost, icebergs, inhumane weather conditions, perpetual darkness in wintertime, and being thousands of kilometers from the rest of civilization does not bode well for job advertisements looking for highly skilled oil engineers and drillers. Tanker crews and pipeline builders also face challenging conditions, which will lead to delays and growing risks of accidents. And if, with the help of fearless venture capitalists, we do get our hands on the Arctic oil reserves, the ramping up of production could take decades. Even building the logistics to the nearest coast that is suitable as an oil port will take years. Roads, housing, and the rest of the mandatory infrastructure must be ready before the actual production in an oil field can start.

Russia is an important oil and natural gas provider for Europe. We will discuss Russia's situation more later, but it is worth mentioning here that the situation is getting more precarious there as well. There is a need for huge new investments in the oilfields of East Siberia and the Arctic sea, since the West Siberian fields face declining production. The infrastructure is largely non-existent and the conditions are challenging for even the best of oil companies. The Russian oil companies have their strengths, but high technology drilling might not be one of them.

Transporting the oil to the market is a problem of its own. Should there be a pipeline, or should the oil be transported with oil tankers? What to do with the pipeline after the production declines? How thick a pipeline should be built? If it is too thin, the field cannot be produced at maximum capacity. If it is too thick, it will be unnecessarily expensive and there will be many problems due to sub-optimal use of pipeline capacity. The construction of the pipeline needs to start years before the oil starts to flow. These huge investment decisions need to be made on a crude estimate of future oil production which is based on just a few preliminary drillings.

With longer transportation routes for oil, there is also the geopolitical perspective to consider. China has been moving in to develop resources in Africa. Can the securing of these transport routes be a partial reason for China's eagerness to build up its naval military capabilities lately?

The expensive lessons from Kashagan

Kazakhstan is located northeast of the Caspian Sea, with China and Russia as neighboring nations. Inside its borders is the Kashagan oil field, discovered in 2000, which was the largest oil discovery in four decades. Twelve years and over 40 billion dollars later, it has become an expensive example of oil production in challenging conditions and the difference between estimates and realities. After massive cost overruns and years of delays, in the summer of 2013, the production facilities finally started up, with production starting by the fall of 2013.[17] It remains to be seen if the current field leases end before the oil companies get their investments back. Goldman Sachs has estimated that Kashagan will need average oil prices of around 125 US$ per barrel to be profitable.[18]

The Caspian Sea is frozen for 5 months of the year, which makes developing the oil field difficult. Temperatures fall below –30°C in the winter, and can be +40 degrees in the summer. The oil is located 4,200 meters beneath the sea floor, under immense pressure. Drilling tends to release toxic and flammable gases. The local government, eager for and dependent on oil revenues, is pressuring the oil companies to start the second phase of the development. Understandably, the oil companies would like to complete the first phase of development first and start the actual production before they make new massive investments into the field. The first phase is estimated to produce 180,000 bpd by the end of 2014, and around 400,000 bpd by 2016. Recoverable reserves have been estimated at around 10–16 billion barrels, which is equal to a maximum of about half a year of global consumption.

While there may someday be a significant production of 1 mbpd from Kashagan, economically it has not been the most promising project so far.

Quality matters

The Arab Spring of 2011 resulted in the civil war in Libya, which also crashed Libyan oil production from almost 2 mbpd to virtually nothing. As the Spring progressed, Saudi Arabia announced that it would release a further 700,000 bpd of oil into the market. The price of oil, that had jumped as the civil war started, did not react much. Why? One of the reasons is that most of the sweet, light crude oil produced in Libya is shipped to Italy for refining in refineries that are equipped to handle this kind of oil. As the Saudis were offering their heavy and sour "bottom of the barrel" oil to the markets, they could not find a refinery that could handle it, and therefore it had little effect on prices. It was a skillful PR stunt from Saudi Arabia, made with little actual increase in production.

The oil sands of Alberta, Canada also deserve a word here, even though they are discussed more later in the book. Instead of pumping, bitumen needs to be mined, and then upgraded so that it can move through pipelines. The whole process is more like manufacturing oil than producing oil. This is also the main reason why it is so slow and expensive to increase its production. The availability of natural gas and sweet water also start to limit the production of bitumen at some point.

Rising oil prices stimulate exploration and the production of oil of lesser quality from more and more distant places. Oil is valuable only after it has been delivered to the markets to be refined. There are a host of obstacles waiting for these new unconventional oils before they reach the gas station, which makes them a weak substitute for the cheap and easy oil we are accustomed to.

Resource nationalism

Resource nationalism is a counter-trend of privatization of assets. It comes in many shapes and forms. It can mean that the development of a new oilfield is entrusted to a new corporation or a joint venture that has a national oil corporation as a major partner. In a more extreme example, it can mean that the lease of an oilfield is first sold to an outside corporation. After the corporation has developed the field and the infrastructure, the production contract is altered by political decisions to make it less favorable to the outside company. It might even mean an outright nationalization of all the facilities and infrastructure. For example, in May 2012 the government of Argentina nationalized YPF, a subsidiary of the Spanish oil company Repsol. The official reason was the failure of Repsol to invest enough in exploration and production. Repsol is suing for $10 billion.[19]

There are a number of negative effects on oil production with these kinds of policies:

- It will be more difficult to get private investments that would make production more effective.
- Other countries might lose trust and impose sanctions that can make it more difficult for the oil from such fields to reach open, international markets (although it has to be admitted that probably the oil will find a buyer nonetheless).

- There may be production declines and a slow process of bringing new fields into production, or mismanaged and even too aggressive production of oilfields that will end up damaging the fields.

Due to the imminent peaking of oil production, it has been increasing in value as a resource. This makes it more and more attractive for governments to nationalize its resources, no matter what the cost. This is easy to justify, since a nation has a right to govern the resources inside its borders, although it might be prudent not to sell those rights first. However, rulers who have meddled too much with the activities of (usually U.S. based) international oil companies have often faced a fatal plane crash or a coup d'état. This may have happened in Iran in the 1950s and in Ecuador and Panama in the 1980s.[20] A topic of lesser attention is the fact that most of the global reserves are owned by national oil companies. The internationals, such as Exxon or Shell, own less than 10 percent of global reserves in total.[21] The production of OPEC countries is mainly under the control of NOCs (national oil companies), and INOCs (international oil companies) are included in projects mainly to bring technology and expertise.

NOCs operate on a different basis from the internationals. This is due to their fundamentally different position in the marketplace. The management of internationals needs to maximize shareholder value and dividends at all times by keeping costs low and profits high. A national oil company does not have similar aims, since their management's main mission is often to keep the local government happy. Governments often have different goals from shareholders. This does not mean that all NOCs are inefficient or badly managed, but it means that they have a different environment in which to work.

On the other hand, this different situation might enable a better consideration of the environment and future generations than the ruthless cost-cutting and profit-maximizing of the internationals. Even so, the most important thing about cutting costs is how and where it is done. If costs are higher, some of the oil will remain uneconomical to produce. It depends on the country as to how well the national oil companies are run. Saudi Arabia for example seems to be managing its oilfields quite proficiently, especially considering the long run and the maximum total oil recovered, despite all the secrecy and the potential for government meddling. The oil-hungry markets find this hard to swallow, since this means that not all oil is produced immediately for them to have as cheaply as possible.

PDVSA

Many nationals are on a tight government leash however. One example is Venezuela. It has the largest oil reserves in the world,[22] and they are held by the national oil company PDVSA. PDVSA was largely controlled by the administration of the late Hugo Chávez. Nearly all of Venezuela's export revenues are from oil, and it is responsible for a third of the GDP. PDVSA is regarded as a swollen, inefficient oil company that is filled with political allies instead of professional management. The company invested 27$ billion into different social programs while it invested 15$ billion in different energy projects. Nobody seems to know what these social programs are and for whom they are meant. The company personnel has doubled since 1999, and debt has grown tenfold since 2006. Production, on the other hand, has decreased by 25 percent since 1999. With accidents and oil spills being common, production is anything but smooth.

The Chávez administration diverted oil revenues mainly to social programs, subventing the price of gasoline (it costs around 3 cents per liter) and food. Nationalizations made a few years ago ran off Exxon and ConocoPhillips. Now their expertise, capital and resources are direly needed to help produce and upgrade the extra heavy oil from the Orinoco Belt.[23]

Rosneft

Another example is Rosneft in Russia. It has an interesting story full of drama and surprises.[24] Rosneft was originally set up as a temporary company, with an agreed lifetime of just 3 years, to take care of some scattered oil production assets of the now defunct Soviet Ministry of Oil until they could be sold or some other use found for them. These assets were mainly some leftovers from the 1992 forming of the three big holding companies Yukos, LUKoil, and Surgutneftegaz, which were to be privatized. With these assets spread all over Russia and a stopgap of 3 years, Rosneft was not a viable private oil company. It was seen more as raw material for creating other oil companies, not as an oil company in itself. Aleksander Putilov, who became the first president of Rosneft, chose not to go by this playbook, but instead ended up keeping Rosneft together as a national oil company for the next 6 years. Rosneft, as planned by Boris Yeltsin's 1992 decree, lost most of its production assets in the first few years, even though Putilov fought back and tried to promote the idea of Rosneft being the national oil company of Russia. For most of the 1990s, it was raided for its best assets by oil rivals, traders, gangsters, politicians, and even its own managers. It is something of a miracle that the company even exists today.

Putilov's successor, Bogdanchikov, continued the battle to keep Rosneft in existence when he took over the company in 1998. Somewhat ironically, he had to fight Putilov, among others, who had carved himself a chunk of ownership in Rosneft's subsidiaries. When Bogdanchikov got Rosneft, the company was practically dead. During the first few years he managed to get the sinking company into better shape. He also managed to somehow convince Moscow that a state-owned oil company was a good idea. The problems, however, did not stop there.

In the early 2000s there was political will, including from the very top politicians such as Vladimir Putin, to absorb Rosneft into the other national energy company Gazprom. This would give Russia a very big energy-champion that would be able to handle both natural gas and oil. Given that the "oil arm" of the new company would be quite small and have little power, and that it would not include the important oil-exporting functions, Bogdanchikov fought back. With some luck, sleight of hand, and bold moves, the merger was cancelled in August 2005. One of those moves was a last-minute bid by a temporary "unknown" company (which was later acquired by non-other than Rosneft) to buy Yuganskneftegaz, an oil-producing enterprise in West Siberia. It was the core asset of Yukos, a privatized Russian oil company which had just been driven to bankruptcy by the Russian government. Yuganskneftegaz remains Rosneft's top producer to date. Rosneft, on the other hand, soon rose to be the top producer in Russia, overtaking LUKoil. After acquiring TNK-BP in 2013, it became the largest publicly traded oil company in the world, but it is still mainly owned by the state.

After 2 decades of privatization, the fossil energy business of Russia is largely back in state control. Russia being Russia, it will be interesting, and of paramount importance, to see how this will play out in the longer term. The key question is whether Rosneft will be able to bring together the best aspects of state ownership and the efficiency and technological competence of an international, openly traded oil company.

The efficiency of oil production

> Even an attempt to get up to 12 mbpd would wreak havoc within a decade by causing damage to the oil fields.
>
> Saudi Aramco official

Technological advancement is often thought to be mankind's best hope to deal with arising problems. A major part of these problems has been solved by developing technology that enables us to use external energy to deal with the problem. The current predicament we are facing is a little different. It has to do with the production and availability of that very same external energy that we have previously used to solve our problems.

In oil production, technological advancement can be roughly divided into two, partly overlapping categories:

1 Development that adds to the total oil reserves we are able to produce. These include deep water drilling rigs, pumping of CO_2 into oilfields to raise pressure, and fracking that includes horizontal drilling and hydraulic fragmentation.
2 Development that makes the production from current oilfields faster. These include for example multipad-drilling and pumping seawater into the oil reservoir.

In addition to technology, the availability of oil is also affected by its market price. If prices go up, a larger part of our oil resources can actually be produced economically. The methods and technologies used depend greatly on the geology of the oilfield in question. Water injection can be used right from the start to add pressure, or at the end of the oilfield's lifetime to make the final squeeze more efficient. In addition to geology, the prices of oil, the method (and therefore price) of production, and the patience of investors all affect decisions on what EOR (enhanced oil recovery) technologies to use.

The second category has to do with our daily oil and sudden supply disruptions. Saudi Arabia has been the world swing producer for decades, increasing or decreasing its production to help stabilize the price of oil. They say that it is not in their interests to have too high a price for oil that will stagnate the economies of their customers and help alternative energy sources to gain momentum. In the light of recent years' events it could be debated whether the Saudis are able or even willing to increase their production to help during a longer supply crisis. Near the end of 2012, Saudi Arabia actually decreased its production in the face of increasing shale oil production from the U.S., to help keep prices high enough.

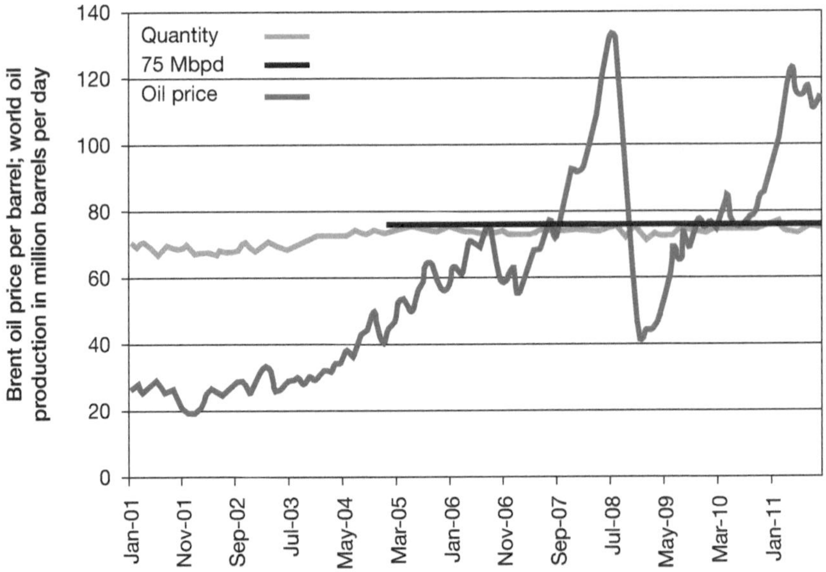

Figure 2.2 Crude oil quantity extracted and price

The production of conventional crude oil correlates poorly with the price of oil, as the production is still at the level of 2005.

Source: http://ourfiniteworld.com/2012/02/01/the-most-important-resource-for-our-future-inexpensive-oil.

Many argue that Saudi Arabia could increase its production from the current 10 mbpd substantially. This would harm the oilfields and therefore reduce their URR (ultimately recoverable resources). It is sensible to keep the oilfields in good shape and therefore maximize long-term production, as the price of oil is not very likely to fall. There are examples of exactly the opposite. In 1980 the OPEC countries increased their production and exports when the price of oil was falling, to compensate for their diminishing export revenues (the price was falling because significant production from Alaska and the North Sea was coming online, causing a glut in the market). The leader of an oil-exporting country, whether dictator or democratically elected, might not have the luxury of thinking decades ahead, if the challenge is to keep the people fed and calm today.

In addition to political limitations, there are geological limitations that can be solved with technology. There are around half a million oil wells in the United States, of which most are producing only a few barrels per day. These are called stripper wells. Huge amounts of water are pumped into these, and the produced oil might be mostly water itself. The oldest still-producing oil well is McClintock No. 1, located in Titusville, Pennsylvania. It has been producing oil for over 150 years, with a current production of 12 barrels per month. The oil is sold as souvenirs in small bottles with a price of 10,000 US$ per barrel.[25]

The reality of oil production is quite different today. The advanced projects use 3D and 4D mapping of the reservoir (4D means that time as well as the movement of oil is taken into consideration). Exxon drilled the record-breaking well, Z-44, on Sakhalin Island in eastern Russia. It was drilled 12,376 meters deep, after which the drill was turned horizontally for another 11,745 meters of drilling to reach the Odoptu field.[26]

Too aggressive production can harm the oilfields permanently. This appears to have happened in Saudi Arabia in the 1970s, in the Soviet Union in the 1980s, and in Iraq in the 2000s. This can also lead to rising costs sooner than usual, as different methods of EOR need to be employed.

The amount of ultimately recoverable oil varies with the technology used and the price of oil, but only to a certain extent. High oil prices may bring new production online, but it will not replace cheap oil. The faster we produce and consume oil today, the faster we run out of cheaply produced oil. Given that oil is a non-renewable resource, the faster we produce it today, and the more aggressively we try to keep production high, the steeper the decline will be when it begins. Big producers such as Saudi Arabia and Russia are likely to be satisfied with current production levels, as they can prolong their production and keep new investments more reasonable. Saudi Arabia has tended its fields rather well. Because of this, and beneficial geology, the average recovery factor of their fields is 56 percent.[27] The global average is around 35 percent.[28]

When considering old oilfields, advanced technology and different methods of EOR have helped to increase their reserves substantially. This growth of reserves during the lifetime of an oilfield tends to be much smaller with newer oilfields. This may be due to the fact that these advanced methods have been available right from the start. This may also be one of the reasons why many optimistic reserve growth estimates have failed; they have presumed that the reserves of new fields will grow similarly to older fields, when in reality this has not happened.[29]

Export Land Model

As the revenues from oil in exporting countries raise their citizens' living standards, they also increase the amount of oil their economies use. This is a significant factor for oil-importing countries that depend on the oil that ends up on the open market. The Export Land Model (ELM), as developed by Jeffrey J. Brown and Samuel Foucher, helps to model and understand this phenomenon.

From the view of oil-importing countries, the preferable model of government in an oil-producing country is a dictatorship as this helps to keep the local population and its energy use in check, and maximizes exports. According to the IEA, by 2020 around 90 percent of the growth in global oil production will come from just five or six countries (Saudi Arabia, Iran, Iraq, Kuwait, and United Arab Emirates). Therefore, keeping the peace in one of these countries can be more important than any experiments with democracy in a dozen small producers. A democracy looking after its people can be a much harder negotiating partner than a friendly dictator. It means that prices are defined in the marketplace, the people involved in the talks can change every few years, and, most importantly, the nations' own oil use will increase if the revenue and standard of living that comes with it are spread more evenly among its population. By using oil and its many refined products, and complaining about the price of gasoline (but failing to mention its origin), we in the developed countries condone all of this. Out of sight, out of mind.

This all goes back a few decades. In 1974, the NSSM200 report, led by Henry Kissinger, reported on how the growing population of developing countries and their growing use of minerals and resources was a threat to the availability of these resources to the United States in the future. The proposed solutions include a family planning program headed by the United Nations to limit population growth. The report also concluded that population growth itself is not the biggest problem, but the industrial development of the countries in question.[30]

Many exporting countries are increasing their domestic oil consumption. If these countries do not similarly increase their production, the amount of exported oil will decrease. Only the oil that ends up in the global marketplace matters for those countries that import oil and buy it from the global markets. Total global production is of secondary concern.

There are many mechanics that affect the domestic oil consumption of producing nations. Population growth is one of the more significant factors. The Saudi population is growing at an annual rate of 8 percent. Oil revenue makes it possible to subsidize food prices, which often (at least initially) leads to population growth, and population growth leads to more energy being used. More than half of the Saudi Arabian population is under 25. In addition, their economy is mainly focused around oil. The oil sector has only a few tens of thousands of jobs to offer, so it is hard to find relevant or fulfilling jobs for all these young adults. This adds to their frustration. It also puts pressure on the government to maintain various social programs to keep them satisfied and calm.

A second important factor in domestic oil consumption growth is rising living standards. A higher standard of living translates into higher consumption of energy. If the country in question is an oil producer, then this energy is often produced with oil. In Saudi Arabia, most of the electricity used is produced with oil and gas (about 50/50). Their domestic oil consumption grew by 6 percent annually between 2004 and 2011. As in western nations, the benefits gained are hard to remove from people without negative political consequences. This means that it is often hard to reduce the consumption of oil, other than by increasing efficiency. Many countries around the

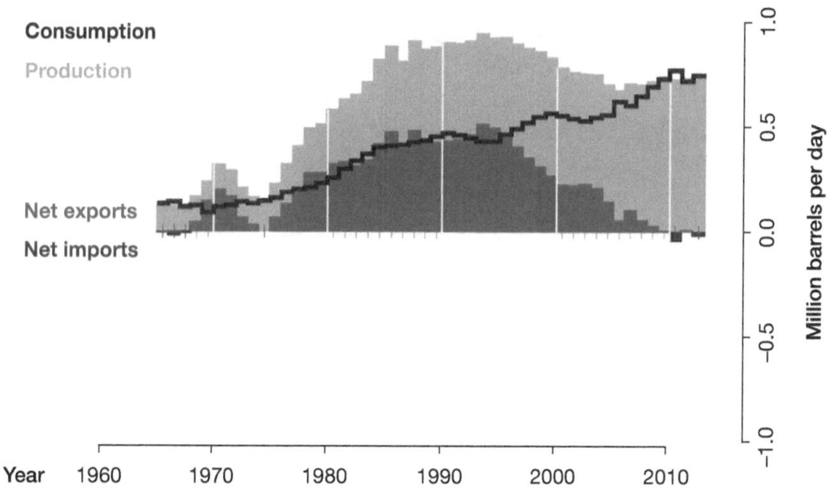

Figure 2.3 Egypt: oil – 2012 imports increased by 282 percent

Showing the radical effect of Egyptian consumption on their oil exports. Even though production is decreasing quite slowly, their exports are collapsing due to increasing domestic consumption. This led to them becoming a net importer in 2011, which incidentally is also the year the people decided to change their government. The civil disorder led to a decline in tourism, which is Egypt's main industry, leading to declining oil use.

Source: BP Statistical Review 2013, http://mazamascience.com/OilExport.

Persian Gulf are planning on reducing their oil use for electricity production, which would leave more oil for export. The nuclear program in Iran might have the most press coverage, but it is not that radical when compared with some other plans in the region. Saudi Arabia is planning massive increases in nuclear power (54 GW), concentrated solar power (25 GW), solar photovoltaic power (16 GW), and wind power (9 GW) by 2050.[31]

If the local government seeks to limit local consumption by some means, for example removing subsidies for gasoline, the result is often social unrest and demonstrations. The people will have their "bread and circuses," and it often translates into imported grain, big cars and cheap gasoline. In Iran the price of gasoline rose by 400 percent in 1 night in December 2010, which led to protests. During the Arab Spring in 2011, the Saudi Arabia government announced that it would spend 130 billion dollars on various social programs to avoid civil unrest.[32] In Nigeria, the cutting of gasoline subsidies led to demonstrations, and the subsidies were partly reinstated. Many oil producers are painting themselves into a corner. Saudi Arabia, for example, needs an oil price of 80–90 dollars per barrel to keep the population happy and the society stable. The same price point for Russia is higher still, estimated at well above 100 dollars per barrel, while Iraq needs around 100 dollars per barrel.[33]

One way to increase exports is the destruction of domestic demand. Military action is one of the more common methods to achieve this. After this, the local government is changed for another, and the rebuilding of lost production capacity is arranged. This has its risks, as Iraq recently has demonstrated. The local population might not like this kind of resource invasion. It can lead to lasting, serious and expensive problems in returning the stability of the local society, which is a prerequisite for increasing oil production. These operations also increase the risk of global terrorism, since an angry and disappointed population offers a good place to find recruits for various organizations. Military action is also damaging to the nation responsible, since it is hard to avoid casualties however superior their military equipment. The costs of equipping, attacking, and then occupying a foreign nation are exceedingly high.

The Export Land Model does not happen to all oil-producing nations. The United Kingdom became an oil importer in 2006, mainly due to rapidly decreasing production from its North Sea fields. The same is true for Norway, which has also seen its production decrease rapidly from its glory days. Many of the bigger, developing oil exporters seem to experience the ELM. These countries have a high demand for oil and the increased living standards it brings.

Forecasts

Daniel Yergin and other analysts at CERA predicted in 2011 that global oil production would grow by 0.9 percent annually for the next 20 years. If we follow the demand trends of 2002 to 2005 of the oil-producing nations, this will mean that exports to the open oil market will decrease. On the other hand, if the oil production predictions of many others come true, with China and India continuing their growth, these exports would disappear completely by 2030. So far, the peak in these net exports has occurred in 2005.

It is clear that these trends cannot continue. The more interesting question is what will turn these trends? And what effects will it have on the global economy, geopolitics, and societies?

Maintaining society and the costs of replacing energy production infrastructure

There are many suggested solutions to mitigate climate change and peak oil. Depending on the background, ideology, and perhaps employer of the person making the suggestions, the solution is that we need simply to change fossil fuel use through energy efficiency, use of renewables, and nuclear power; we should change from cars using internal combustion engines to buses, hydrogen cars, electric vehicles, or even cars that use boron; and while we are at it, we should change the whole hydrocarbon-based economy into a hydrogen economy.

These are all sound ideas, at least on paper and in speeches. In reality, the picture changes dramatically. If we aim to rebuild our energy production on a massive scale or change our logistical infrastructure and fleet, it might prove to be politically very unpopular and operationally difficult. One of the key problems is the vast difference in the timescales of building energy infrastructure and daily politics. Politics are usually done on a timescale of a few years, due to election cycles. This means that hard and unpopular decisions in the long run become next to impossible, since the politicians want to get re-elected. Building power plants and energy infrastructure has a timescale of several decades. A power plant built today is likely to be operational for 40–60 years. The result of this difference can be observed in many nations' public debates on energy and energy policy. Depending on the ruling party, this or that energy source is taxed or subsidized more or less. This often results in a complex and unstable investment environment for energy producers.

Alternative energy sources are more upfront as investments, in terms of both money and energy, than conventional, fossil-fueled power plants. This means that most of the investment is in building the alternative power plant, be it a nuclear power plant or a windmill, and relatively little is paid for the fuel, maintenance, and such. So far this has not been a big problem, since any relatively sensible and viable energy investment has had both funding and energy available. If these become more limited, or even start to decline as fossil fuels deplete, societies will need to decide where these diminishing resources will be allocated: consumption or investments. Enter the energy trap.

The energy trap

What is the energy trap? We can try to figure it out with a simplified example. In reality, it is not just about energy, but also other raw materials, rare earth minerals and other goods that require energy in their production and refinement and are limited in their production volume.
Basic assumptions:

- The "power plant" to be built has an EROEI average of 15:1. This means it will produce 15 times the energy during its lifetime than went into building it.
- The power plant's lifetime is 30 years (so the power plant will take 2 years to produce the amount of energy that went into its building).
- Fossil fuel production decreases by 2 percentage points per year.

In option one, the nation will not build more alternative power plants. The mainly fossil energy used for upholding its consumption and living standards will diminish by 2 percentage points annually. This alone is sufficient to cause economic problems that keep getting worse every year.

In option two, the nation starts a program to replace its diminishing fossil-fuel based energy production capacity with alternative energy production every year. Instead of losing just 2 percentage points per year from consumption, the nation must divert a further 4 percentage points to building alternative energy, which will only start producing after the investment is made. So the nation and its people will have 6 percent less energy and resources for consumption in the first year. And as the next year sees another 2 percent drop in fossil-fuel based energy, the building project needs to continue. Actually, it needs to continue for decades, and will eat a big portion of the nation's available resources during that time.

The nuclear debate suffers from the energy trap already in many countries. Nuclear power plants are seen as too expensive investments that need to be made largely upfront, and the payback times are seen as too long for investors. Practically speaking, this means that long-term investment is left in the shadow of today's consumption.

The energy trap is especially hard for politicians and other decision makers, due to the shorter time span of politics compared with the energy industry. If our available energy started to decrease by 2 percent annually, politicians would likely have their hands full coping with the associated negative effects, let alone that they would have to explain to their populace that actually they needed to consume far less, for decades, for there to be sufficient resources to ramp up alternative energy sources.

What about increasing efficiency? From the viewpoint of the energy trap, investment made to increase energy efficiency in different processes is similar to building new capacity. The investment is made largely upfront, and it will take years for it to pay back. The same goes for vehicle fleets, buildings and restructuring communities through better zoning and so on. It will be very hard to replace infrastructure before its operational lifespan is over. Furthermore, Jevons' Paradox, or the rebound effect as it is also called, will see to it that increases in efficiency lead to more energy and resources being consumed, at least partly. Cheaper goods and fuels will usually push for more consumption, which will often lead to more resources being used at the end of the day.

A further problem arises from the fact that a lot of the invested energy is oil, but usually power plants produce mainly electricity (and process heat). On a global level this leads to a zero sum game, where the energy (oil) used is always reduced from somewhere else. Who will decide, on what basis and on whose authority, where this oil will be taken from? And what will the truck driver do with the electricity he gets in the following years, assuming his company is still in business after its personal oil crisis?

The energy trap is further discussed in the popular "Do the Math" blog.[34] Some solutions to it include:

- starting the shift to alternative energy early enough (before depletion sets in and prices skyrocket);
- planning and making energy policy with a long-term perspective;
- reducing energy use systematically and in a controlled manner;
- applying tight efficiency standards on investments in the long term.

Diminishing usable energy (exergy)

According to the first law of thermodynamics, energy cannot be created or destroyed. However, the usable energy that is in an easily usable form, *exergy*, is depleting.[35] When we burn a liter of gasoline in the car, the energy in it does not disappear; but the easy-to-use energy embedded in the gasoline is transferred to movement and eventually to waste heat that cannot be used again for anything.

On a global level, more than 80 percent of our primary energy comes from fossil fuels. As a rule of thumb, all these fuels have a weakening net energy in the long run, since we use the easiest and richest sources first. This means that more effort (spent energy) is needed to produce an increasing amount of energy and resources. At the same time, around 200,000 new users for this energy are born every day. The situation where less and less net energy is left for the society outside energy production is not far. This situation is also a new one for most of us living in the developed world, since we have grown used to having more and better every day.

It will become problematic as we need to take more and more of this net energy away from consumption and other investments, and direct it into alternative energy investments. The biggest questions are how soon this depletion of net energy will start and how quickly it will get worse. The operational lifetime of different types of power plants and their EROEI play a key role in what will be sensible. One further question is our intellectual, social and emotional readiness to accept the new situation and what it means for many of us.

Peak oil perspective

Electricity can easily be made without oil, but replacing oil is harder. Peak oil, on the other hand, is first a liquid fuels crisis, and only later an energy crisis. The "oil trap" is even harder to solve, since alternative fuels are hard and expensive to make or synthesize, and the efficiency of their production is often poor.

If we aim to change the power source for our logistics, we are faced with huge investments filled with obstacles. The production of electric cars cannot ramp up very fast. Alternative fuels lack distribution infrastructure – it is hard to produce, transport, and store hydrogen, for example. Moving traffic to rails needs big projects that can take years, even decades, and it is only economical with enough passengers and for certain routes. The fuel efficiency of passenger cars has been improving thanks to regulation and taxation, but the efficiency of the total fleet improves only slowly. In addition, these improvements get harder and harder. We probably won't get very near to zero litres per 100 km.

One more thing to understand is that we cannot afford to build many different infrastructures in the same place, be it for traffic or distributing fuels and energy. All of these will require maintenance. What we need is a holistic view on directing resources and coming up with solutions. Since the marketplace seems to be largely blind to the scale of the coming problem and its schedule, the invisible hand of the marketplace is inefficiently doing this or that here and there. At the same time, different interest groups distort decision making and the information that guides the marketplace for their own benefit. The only thing that seems sure is that blame for any problems will be found somewhere else.

Even were we to discover a totally clean and harmless energy source tomorrow, the situation would be unlikely to improve much. The establishment benefits from the current situation immensely, so they would fight tooth and nail against any competitive technologies. This would, due to their substantial resources and wealth, prolong the transition so that they would have time to suck every dime from the current system.

We have built massive infrastructure during the last century or so. It requires huge ongoing investments just to maintain and repair it. With climate change and worsening environmental catastrophes, this maintenance will only get more expensive. The more developed and complex our society is, the more prone it is to natural disasters and the damage they cause. Repairing the damage caused by a natural disaster might show positively in a national GDP, but this has nothing to do with raising people's living standards in the long run.

Economic limitations and motives

The free market is often described as the invisible hand that increases or decreases demand and supply automatically as they are needed for maximum societal benefit. When we discuss oil, however, the invisible hand belongs to those that benefit from the current situation. Its main job is to shift wealth from others into the pockets of its owners, not caring much about the societal benefits.

If peak oil and its long-term effects had truly been taken into account in the marketplace, the price of oil would likely have risen far above current levels. This would have led to decreased demand (and supply) before we faced geological and technological limits. This has not happened. Why?

One reason is the limit and brevity of the time perspective. The demand and supply of oil is decided by politicians on the timescale of election cycles; the companies and their CEOs living in the quarterly economy and annual goals; and the consumers operating according to the timescale of their next paycheck or reality TV show. Another reason is economic benefit. The people and organizations that have collected wealth and power from oil profits will most likely want to hold on to these, and it is easiest to do so by prolonging the current situation.

The third reason is the ease of oil as the source of wealth. It is needed everywhere, and all the time in greater quantities. Unlike agriculture, the weather does not play as big a part in oil production. The producer has no need to develop a new version of oil every year to compete with other producers. All that has been done is that a decision has been made to produce oil, and soon afterwards money has started to flow into their pockets. For these reasons it has been beneficial to produce oil like there is no tomorrow, until it finally starts to seem like there won't be one.

Highest price

Traditionally, most oil producers have maximized their production, and the price has been set at the lowest-selling barrel. When a commodity such as oil cannot meet demand with supply, the price will instead be set by the highest bidder. As the price increases, it encourages more production.

Oil exporters have invested a substantial amount and amassed great wealth in recent decades. If the expected profits from these investments, both old and new, start to be smaller than the expected rise in the price of oil, it will be economically sensible and logical to leave part of the oil underground for now. Since the price of oil is rising, more wealth will be accumulated anyway. All of this leads to the possibility of production dropping faster than it would drop for purely geological reasons.

Another economical limitation is from the demand side. It seems that there is a maximum price the developed countries can pay for their imported oil. After that price is exceeded, their economies will go into recession, decreasing their demand for oil. In this sense, the demand for oil does have some price elasticity, but only at the expense of economic contraction. This is a fundamental argument of this book; we should aim to reduce the negative effects of peak oil in some other, less destructive ways than going into a recession every time the price of oil goes above a threshold.

In 2008 the threshold was around 147 dollars per barrel. The massive bursting of real estate and derivative bubbles blown by the financial industry took attention away from oil prices. But many analysts and experts have said that the high oil prices functioned as the needle that ended up bursting the bubble. In 2001 and 2012, as the EU was suffering with its debt crisis, that threshold seemed to be around 120 dollars per barrel. Since the Euro was weakened substantially by February 2012, the price of imported oil (after buying more expensive dollars to pay for it) was even higher for Europeans than in 2008.

Those oil reserves that are only economical to produce with prices upwards of 150 dollars could run into a world that is unable to pay that kind of price. This means that they will, for the time being, stay underground, where they will await the time when humanity will have made its oil use more efficient and developed cheaper methods of production. As an example, the future of production in Alaska's North Slope is largely dependent on future oil prices. If the price were to fall below 62 dollars per barrel, the production would not be economical by 2025. This is because the pipeline needs a large enough flow to operate properly, but the development of new fields at that price will not be economical. This would lead to a rapid decline and abandonment of current fields. At a higher price, more along the lines of 200 dollars per barrel, the output of Alaska would grow substantially.[36]

Last, sustained high oil prices will lead to a rise in break-even prices[37] in oil-producing countries. This is due to rising living standards and energy use that will result from higher export revenues. The price area where both the global economy and the societies of oil producers can function without major problems, is getting smaller every day.

Growing population of oil users

In the early 1980s, the subject of peak oil was completely in the margins. The world, nonetheless, went through an important peak oil event without anyone really noticing. The peak was oil being produced per person at 5.3 barrels per person per year. Since then, the population has grown more rapidly than oil production, and the amount is now around 4.4 barrels per person. The vast majority of oil is consumed in rich, developed nations, which leaves less for the citizens of poorer countries. Canadians consume around 20 barrels per person per year, whereas people in India get by with around 1 barrel per person per year. In terms of liters per day, Canadians consume around 10 and Indians around 0.5, while the Europeans consume around 5 liters per day. This situation is now changing. For the people of developed nations who are used to consuming much more oil, this is grave news.

The relationship between oil and the economy has often been described as follows:

1 Economic growth increases demand for oil.
2 Supply cannot meet this higher demand and the price of oil rises.
3 Rising oil prices raise the cost of living and basic necessities, which decreases discretionary spending.
4 This leads to economic recession or slower growth, which leads to lower demand and prices for oil.
5 A lower oil price makes economic growth easier. Return to part one.

For one reason or another, many recent recessions have been preceded by a hike in the price of oil. It seems that cheap and plentiful oil makes economic growth possible if all the other factors remain in place. However, expensive and scarce oil prevents economic growth even if the other factors are in place.

As the Arab Spring movement reached Libya, it led to events that practically closed down the country's 2 mbpd of oil production. This led to oil prices rapidly rising to 120 dollars per barrel, and many developed nations went into recession, threatening the global economy. But why did the prices of oil stay high despite this? The short answer is China, India and the other BRICS countries (Brazil, Russia, and South Africa).

There are hundreds of millions of people moving up and into the middle class in these countries. Many of them dream of their first automobile. India produces 1.8 times more commodities for its consumers per barrel of oil used, compared with the United States. For China, this multiplier is 1.5.[38] China and India have bought all the oil that the developed nations could not afford in recent years. This demand has kept the price of oil high, and therefore the recession continues for the time being. China's oil use is growing in lockstep with its economy, at around 7 percent per year. If this trend continues, oil consumption in China could double in 10 years. China became a net importer of oil in 1992, and currently imports more than half of its oil.[39] Chinese demand will account for more than 60 percent of demand growth between 2008 and 2035.[40]

Permanently expensive oil?

As oil becomes permanently expensive, there will be a few consequences:

- it will increase oil exploration;
- it will increase the development and use of different replacements for oil;
- it will increase the efficiency of our oil use.

In addition, it will make significant economic growth in the developed countries impossible. This will lead to further problems. Debt crises, high unemployment, civil unrest and political instability might become more permanent realities among western nations. Expensive oil and economic difficulties will make building a more oil-efficient infrastructure relatively more expensive, and therefore politically harder to implement.

As the situation progresses, China and other export-driven developing countries are starting to see problems of their own. This is an indirect effect that results in decreasing their exports to Europe and the United States, which cannot afford to import as many goods as before. As transportation costs become a bigger part of a product's price, we will see a shift to more localized production. Globalization and global trade are forces that bring the world closer together, but might reverse or at least slow down considerably. The geopolitical implications of this trend are unpredictable.

As production stagnates, a growing base of oil users will mean that there will be less for everyone. This trend of less oil per person has existed for more than three decades. But now the oil-using middle class is growing rapidly. This could lead to less open markets and more bilateral trade between political allies. As there is not enough oil to go around, the price will rise and someone will have to do without. Even the richer nations in Europe, such as Germany, could find themselves in a situation where their European neighbors can no longer afford to import their products. Diminishing exports will lead to a diminishing ability to pay for imported oil.

Consequences of oil crises

The oil crisis from October 1973 to March 1974 was caused by an oil embargo run by OPEC countries. It was a counter to U.S. support for Israel in the Yom Kippur war. At the beginning of the crisis, OPEC announced that it would increase its oil prices by 70 percent, to 5.11 dollars per barrel. In addition, many OPEC countries embargoed the U.S. and its allies, such as the Netherlands, from their oil exports. In 3 months the price of oil rose from 3 dollars to 12 dollars per barrel (approximately 40 dollars in 2008 currency).

The defeat in Vietnam and the Watergate scandal that took place around the same time eroded people's trust in U.S. global supremacy. After 2 decades of strong growth, the average paycheck started to shrink. The economic competition in the global markets had been getting tighter, and U.S. producers saw a 30 percent drop in their profits. The sudden peaking in domestic oil production in 1971, and the decline thereafter were also a strain on the U.S. economy. With all this in the background, the oil embargo and the possibility of future supply problems was enough to instill panic within people.

Suddenly, everybody wanted to have their gasoline tanks full to the brim. This led to growing queues at gas stations, which led to eventual real, if local, supply problems. This fed the flames of panic further. The crisis was made worse by the government, as it started to regulate oil prices. It set a ceiling on the price of already-producing "old oil," and left the "new" production without such a roof in an effort to boost new production. This led to the prompt removal of even the old oil production, worsening the crisis. A speed limit of 88 km/h (60 mph) was imposed, along with many other limits and regulations. These often just led to longer queues and worsening problems at the gasoline pump.

The U.S. also started to investigate the possibility of improving the fuel efficiency of personal vehicles through legislation. This led to the Corporate Average Fuel Economy (CAFE) standard later in 1975. Oil use was also reduced in electricity and residential heating. Many western nations stepped up their oil exploration efforts. This led to significant discoveries in the North Sea, for example. Energy conservation became the buzzword of the day, and alternative energy sources such as natural gas, peat and nuclear power saw a rapid ramping up. It was not all an environmental awakening however, as there were serious efforts to produce kerogen from oil shale (more on kerogen later in the book).

The OECD organization revealed its inability to respond to such a crisis. Therefore, as suggested by Henry Kissinger, the International Energy Agency (IEA) was founded in November 1974. Its mission was to enable a coherent, planned response to oil market disruptions by its members, and to collect and report global energy statistics. This also led to the building of strategic oil reserves in OECD countries (of which almost all were also IEA members). Since their inception, these reserves have seen three occasions where oil has been released from them: during the war in the Persian Gulf in 1991, after hurricane Katrina in 2005, and during the 2011 Arab Spring.

Due to the 1973 oil crisis, inflation skyrocketed and economies plummeted in many western countries. This caused civil unrest everywhere. Both the extreme left and the extreme right saw their political power grow in Europe, as they tapped into the growing dissatisfaction of the population during the crisis. Many nations installed various states of emergency, limiting oil use in heating, electricity and transportation. Since oil was used in electricity production, large savings could be had by limiting electricity use, for example by lowering temperatures in people's homes. Another way was to lower the speed limits on highways, which was also implemented in many countries and still is among IEA responses to oil supply disruptions. For example, Sweden limited the buying of liquid fuels to a maximum amount of 100 liters per 7 weeks.

The IEA also has a variety of softer methods to help lower oil consumption. Many of them would be sensible even without a sudden crisis. These include ride-sharing and avoiding energy use that is not essential. So far, the desire for maximum comfort and the "good life" rules among most, and these methods are left to political speeches and plans for the (distant) future.

The 1979 energy crisis

As the 1970s went on, oil prices kept creeping upwards and stock exchanges kept crashing. The OPEC countries refused to lower the price of oil, and it hovered at around 40 US$ per barrel (2008 dollars). As their oil wealth accumulated, many OPEC countries, Saudi Arabia among the first, started planning the nationalization of their oil reserves and companies. The second abrupt crisis started as Iran went through a revolution and its oil production collapsed. Other OPEC countries such as Saudi Arabia increased their oil production to help reduce the fall in oil supply, which ended up at around 4 percent. Iran's new government managed to get the oil production going again, but it never regained its previous levels. The panic started to spread again. In 1980 Iraq, supported by other Arab nations and the United States, invaded Iran. This resulted in a near total collapse of their oil production, with Iraq's production also suffering. By 1982 Iran had reconquered their lost areas, but the war lasted for 6 more years, taking a heavy toll on both its people and infrastructure. It is estimated that around half a million people lost their lives. Borders were not moved, and war reparations were not paid.

The Iranian revolution led to a doubling of oil prices, which rose to around 100 dollars per barrel. The inflation reached its highest level in 1980 at 13.5 percent. Many remembered well the panic of 6 years before, and the hoarding started again. According to some estimates,[41] around 150,000 barrels of oil were used daily just by idling in queues at the gas stations. Oil companies were accused of hoarding and driving prices up, but in reality there was only a 3.5 percent drop in petroleum sales from the previous record year.

With President Carter at the helm, the government tried to persuade people to conserve energy once more. Carter had solar water heaters installed at the White House, which were later removed by the Reagan administration. Carter was also responsible for founding the Department of Energy, which Reagan also tried to shut down later (without success). Carter also established the Carter Doctrine, which stated that the United States would not tolerate any threats to its vital interests in the Persian Gulf.[42]

The first golden age of nuclear power started to lessen the demand for oil in electricity production, as did a shift to more coal being produced. In 1980, global oil production started to grow again, as the fields from Alaska and North Sea came into production. By mid-decade the price of oil had crashed, and it continued crashing well into the 1990s. As prices plummeted, so disappeared any worries about the abundance of oil. As the Soviet Union crashed a few years later, losing 4 mbpd of its production, there was no major oil crisis, as there was a lot of spare capacity waiting to be turned on.

Fragile society

The progress of societies is usually seen in a positive light. The 2000 events in England cast a different light on societal progress, towards more fragility.

The whole chain of events started as the price of gasoline rose above the psychological barrier of £4 per gallon. It began as a spontaneous protest by 150 farmers, truckers and taxi drivers that started to gain momentum through radio and cell phones around the country. Demonstrators ended up blockading oil refineries and production facilities so that instead of the normal twenty trucks per hour, only one truck managed to get through on average.

The oil companies seized the moment to extort tax relief from the government and dragged their feet in starting the deliveries again. The government was caught with its pants down, completely surprised. After 5 days, a third of the gas stations were closed as tanks ran dry. Ambulances had fuel left for just 2 days. The sixth day saw fights erupting at gas stations as tensions ran high. Gasoline was stolen even from cars in the parking lots of hospitals, which made it hard for hospital personnel to get to work. Auto manufacturers announced that they would have to stop running their factories in 2 days. Big retailers started rationing bread and milk. Supermarkets announced that they would have fuel left for 12–48 hours of deliveries. Our fine-tuned Just-In-Time society faced all of this in just under a week, even if there was no actual shortage of oil, just wide disruptions in its delivery.

Even a rumor of supply disruptions can drive people to panic, which spreads fast. For most, the first thought is to go and fill up the car. The stocks at gas stations and the fleet of delivery companies are rated for normal operations, which means that an average car will have its tank around half full. If most car owners suddenly decided to fill up, gas stations would soon have long queues. As the idling at gas stations gets worse, rumors of dwindling supply stocks start to spread, along with some gas stations actually running out of fuel. This feeds the restlessness and panic further, activating a positive, self-enforcing feedback loop that makes the actual problems far worse than they would have otherwise been. Security analysts worry about this feedback loop if the scale turns from local to national and global. This would have countries themselves panicking and starting to hoard oil in fear of a global supply disruption. This could also lead to military action.[43]

So what can we learn from the oil crises of the 1970s? Well it is clear that we have not learned much, and even those old lessons are limited in value for application to the current world. The warnings of permanently expensive oil felt like old overreacting as the oil price plummeted in the 1980s. Back then, oil was used in electricity production, which is not the case any more. We can no longer save much oil by saving electricity. The cuts will have to happen elsewhere, and they will be more painful. Back then the oil crises were temporary, whereas peak oil can plague us with more lasting problems.

The supply crunch is erroneously viewed as a temporary problem that can be solved by releasing oil from strategic reserves or by using technology.

The need to accustom ourselves to the new situation has not yet been realized, and therefore no meaningful public dialog can be had.

Compared with the previous oil crises in the 1970s, the following things at least have changed:

1. The threat is no longer a temporary oil embargo, after which the oil starts to flow again, but a permanent depletion of a finite natural resource.
2. There are not likely to be discoveries like the North Sea to save the day.
3. Spare capacity is getting tighter, which will make the response to crises harder.
4. Oil consumption in OPEC countries is increasing rapidly, leading to diminishing exports.
5. As shown by recent events of the prolonged Arab Spring, the population explosion of the past 3 decades has led to millions of idle young men. With rising oil and food prices, this is a hazardous situation that can threaten even current production capacity.

There have also been improvements. The strategic reserves, a minimum of 90 days of oil net imports or 61 days of consumption, held by European Union members help to stabilize the markets just by existing. The oil efficiency of western countries' economies has improved greatly during the last 3 decades. We get much more done with a barrel of oil than we did in the 1970s. But our dependency on oil has not decreased, since nearly all transportation is directly powered by oil. As the events of 2000 in the UK showed us, our current industrialized, urbanized, and increasingly complex societies are as dependent as ever on the constant flow of oil. Maybe even more so.

Political instability in oil-producing nations and the stretched U.S. military might

> Let our position be absolutely clear: An attempt by any outside force to gain control of the Persian Gulf region will be regarded as an assault on the vital interests of the United States of America, and such an assault will be repelled by any means necessary, including military force.
>
> President Jimmy Carter, 1980

Recent dreams of a shale-oil powered oil independence aside; a constant headache for the U.S. government has been their reliance on imported oil. A large part of these imports come from politically unstable areas such as the Middle East and Africa. On the other hand, many citizens in the oil-producing nations feel that it is unfair that outsiders use their oil to run western, rich societies. This provides fertile ground for local conflicts and unrest, and also for terrorist recruitment (even if they often get their funding at least indirectly from oil money).

Saudi Arabia

Saudi Arabia may be the most important oil producer. It produces over 10 percent of global oil and has between 15 and 25 percent of proven reserves, depending on who you ask. It is ruled by the Saudi family that managed to gather and unite the small, constantly fighting tribes together as a nation by the turn of the twentieth century. The timing was good, as, soon after, oil was found on the coast of the Persian Gulf. One might argue that the power of the ruling family is based on two things. First is their ability to content the people with oil money and lavish social programs. Second is the private army that has been equipped with the same oil money, which creates a credible threat of violence. As the Arab Spring of 2011 threatened to spread into Saudi Arabia, the government managed to keep it under control by offering more social benefits, pleading to the citizens' faith and religion, and with the threat of violence. Such social benefits had their price, as they put an effective floor on the oil prices that Saudi Arabia needed for its oil. This floor was reported to be in the vicinity of 100 dollars per barrel.[44] Drawing on these same oil revenues is also the private army of the ruling family, as well as the thousands of family members and princes with their often lavish lifestyles, which has a price tag of around 10 US$ billion per year.[45]

Saudi Arabia has a population of around 24 million, of which around 5 million are immigrant workers. This equals around 80 percent of the working force. The oil sector is responsible for over half of the nation's GDP, and 90 percent of export revenue. The national oil company, Saudi Aramco, employs roughly 50,000 employees. The demographics are

Grand Mosque Seizure

On 20th March 1979 a group of Islamist dissidents attacked and seized control of Islam's holiest place, the Grand Mosque of Mecca. Their demands included an oil embargo against the United States, and the deportation of foreign civilian and military personnel from the Arabian Peninsula. The seizure was led by Mohammed Abdullah al-Qahtani, declared as the Mahdi, the redeemer of Islam. The siege lasted for 2 weeks, during which time hundreds of civilians and militants were killed. As a result, the Saudi state implemented a stricter enforcement of Islamic code.

worrying: 32 percent of the population is under 15 years old, and 78 percent are under 40 years old.[46] When this is combined with Saudi Arabia's 11 percent unemployment rate, it amounts to a large population of idle young men. This has often been seen as a recipe for societal upheaval and a source of unrest.

Even if the collapse of Saudi rule seems improbable, the risks are growing with the problems. The consequences to the political stability of the Middle East and to the global oil production would be huge and hard to predict. In addition to Saudi Arabia, there are a number of other, smaller but unstable oil nations.

A stretched U.S. military machine

The vastly superior military might, both in quality and quantity, of the United States functions as a global peace-keeping police for both the oil exporters and oil importers. The United States has hundreds of military bases all around the world. They ensure that it gets the benefits and responsibilities of the office of global police. It is a rare topic among Europeans that we enjoy security services worth billions every year without paying much for it ourselves. We enjoy those benefits every time we start our car or do global trade. It is hard to get exact data on the oil use of the U.S. military. One estimate is that it consumed around 400,000 barrels of oil a day in the recent Iraq war. It is trying to ensure that it gets the daily, domestic oil use of around 300,000 bpd from liquefied coal, if needed.[47] The Pentagon is the largest single oil buyer in the world. President George W. Bush has openly admitted that the war had much to do with the oil in the region.[48] The same has been said by the U.S. Federal Reserve's former chairman Alan Greenspan, when he wrote in his memoirs: "I am saddened that it is politically inconvenient to acknowledge what everyone knows: the Iraq war is largely about oil."[49]

The military machine of the United States has started to be stretched thin, due to its immense budget and the ever-growing federal debt, especially in the 2000s. It seems that the United States is becoming less and less willing to smother the local conflicts around the world with its military boots. The Vietnam War and the Iraq operation taught that there are no easy and quick victories to be had, no matter how superior one's military might may be. It may not be willing to tie its units for years to come in every conflict that erupts. This lack of willingness was demonstrated during the Libyan revolution, where the leadership was quickly shifted to NATO and therefore to the European Union. One of the reasons might be the fact that the European Union is the main importer of Libyan oil.

The United States also announced in late 2011 that it would shift the focus of its troops from Europe to the Pacific Ocean to counter the growing military might of China. This means that Europe will have to see to its own security more in the future. Depending on one's viewpoint, this can be seen as a good or a bad thing.

The vulnerability of oil refineries and infrastructure

> I recently participated in Beijing in what was called a 'petroleum disruption exercise.' We posited a terrorist attack against Saudi production, for example. We had shippers and insurers participating in the exercise – Lloyds of London and so forth – and the price of oil went out of sight. Four hundred dollars a barrel. Shippers wouldn't ship, and insurers wouldn't insure.
>
> Colonel Lawrence Wilkerson, 2012

In addition to political instability, the infrastructure of producing, refining and transporting oil is rather vulnerable. Since the fine-tuned global economy erupts into an epileptic seizure at the slightest disruption, the oil-fueled machine that keeps it humming and spinning must operate like a dream. Delivering 90 mbpd, more than 1,000 barrels per second, calls for a flawless operation of an impressive industry and infrastructure.

There are a lot of instances and points in that infrastructure that could lead to relatively huge disruption compared with the effort needed. This makes oil infrastructure a tempting target for extremists, terrorists, and others who, for example, see the western way of life as an abomination.

Pipelines

Oil pipelines offer thousands of kilometers of potential targets that can be practically impossible to guard effectively. The majority run below ground, which makes sabotage a little more difficult, but by no means impossible. Shovels and explosives are cheap. Pipelines also run through many nations, which creates an incentive for those countries to use them to further their political goals.

In March 2012 a report spread through Twitter, news blogs, and the Iranian state news channel, that one of Saudi Arabia's oil pipelines had been damaged in an explosion. The price of oil immediately jumped by 5 percent as Brent hit 128.4 dollars per barrel. Afterwards, the Saudi government denied the rumor and the price decreased.

In October 2012, PPK, a Kurdish organization, blew up a gas pipe between Iran and Turkey. There are a number of oil and gas pipelines built and planned to go through Turkey from the Middle East to the European markets. In February 2013, a bomb attack disabled a 16-inch fuel oil pipeline for several days in a worrying sign of growing political unrest in Iraq.[50]

Another potential threat to pipelines is offered by climate change. If the permafrost in Siberia and Alaska starts to melt, the oil and gas pipelines built there could face hard conditions.

Refineries

Refineries are another vulnerable target in the oil infrastructure. Petroleum products are easily transported, so disturbances in a local refinery often cause only temporary and local effects. On the other hand, many refineries are often situated in the same area, making it a hub of oil transportation. This makes them an interesting target for those who seek to disturb the stability of the global economy or the oil-dependent western world.

Refineries are also prone to cyber-attacks that have broader aims. In April 2012, apparently the United States or Israel attacked the Iranian oil ministry with a virus that forced Iran to unplug six refineries from the information network to prevent the virus from spreading to them.[51] In October 2012 another virus wiped all files from 30,000 Saudi Aramco computers, making them useless, in an attack that was described as the most sophisticated attack seen in the business sector.[52]

In February 2011, two armed men attacked the Baiji refinery, located north of Baghdad, and managed to detonate several bombs.[53] The output of the refinery was cut in half and the repairs lasted for a month and a half.

Oil ports and other important transportation routes

In 2011, 35 percent of oil was transported by tanker and 20 percent of all oil on the market was transported on supertankers through the Strait of Hormuz. A history of restlessness and war in the area has made it almost impossible to build pipelines, so the oil has to be transported by tankers. There are only a couple of thin pipelines to the Mediterranean Sea, and only recently have some pipelines been built to the Indian Ocean. As tensions rose with the Iranian nuclear weapons program, it resulted in higher oil prices. This was because Iran was seen to have a credible capability to close the Strait of Hormuz, if, for example, Israel mounted a pre-emptive strike against Iran's nuclear facilities. There are seven important straits in the world, of which Hormuz and Malacca (between the Indian Ocean and the Pacific Ocean) are the most crucial.[54]

Colonel Lawrence Wilkerson, quoted above in an interview with *Vanity Fair*,[55] was very worried about the situation escalating in the Persian Gulf. According to him, there are many participants outside the U.S. military that see immense profits in the possibility of war in the area (oil companies and military equipment manufacturers), or the possibility of replacing the Iranian government with another (such as neo-conservative politicians and their backers). In fact, it is the Pentagon that seems to be the most averse to an armed conflict in the area, given that it is perhaps the most aware of the likely consequences.

Even though the IEA mandates strategic reserves for its members, it is worth remembering that crises can stretch on for years, as happened in Iraq. As crises go on, so also does the escalation of oil prices. Replenishing the strategic reserves with much higher oil prices later will be very costly.

According to Colonel Wilkerson a pre-emptive strike would lead to chaos in the region. War, terrorism, oil delivery disruptions, and conflict that could stretch for decades would leave the U.S. with no chance for withdrawal. The global economy would go into a tailspin as prices of oil skyrocketed and supply became unreliable. This could lead to unforeseeable consequences in global geopolitics. Wilkerson even sees Iran with nuclear capabilities as a better option. It would likely lead to Saudi Arabia buying a nuclear weapon from Pakistan, which would lead to a stabilizing stalemate in the region – a deterrent.

Political and paper oil reserves

Peak oil dynamics is made worse by the fact that there is no reliable information on global oil reserves. Given how dependent we are on oil, it is hard to imagine that we have no way of getting credible, third party estimates on how much oil we have left. What we have to make do with are the one-sided, unverified announcements made by oil-producing nations.

Approximately 250–300 billion barrels of global reserves are based on announcements made by OPEC countries made between 1985 and 1990. Since there were no major exploration efforts going on, these reserves are often called *political oil reserves*. In 1985, Kuwait increased its reserves suddenly by 40 percent. Three years later, Venezuela decided to include the extra heavy oil from the Orinoco Belt into its reserves, effectively doubling them. Other OPEC countries followed: Abu Dhabi +197 percent, Dubai +186 percent, Iran +90 percent, Iraq +112 percent, and Saudi Arabia +51 percent.[56] It is hardly credible that three nations, Iraq, the UAE and Kuwait, would increase their reserves to the same level inside a few years. In addition, none of those reserves have diminished, or even changed, even though immense amounts of oil have been produced in the span of over 2 decades. They have replenished their reserves with the exact same amounts they have produced.

It may not be a coincidence that in the 1980s OPEC was planning on production quotas for its members that would be based on each country's remaining reserves. The more reserves a country had, the bigger its production quota, and the larger the revenues for the local government. As one country started raising the stakes, others had to follow. Without much exploration or investment, the global oil reserves grew to new heights in just a few years – very convenient.

OPEC has never been a very united group. The members do not trust each other's reserve estimates, so information gathering is done by a third party. In the not-so-distant past, oil production estimates were done by such scientific means as planting a guy with binoculars near an oil port to record how much draft the tankers had.

To make the big picture even foggier, OPEC countries will not let any third party estimate their reserves, and are very secretive about their production information. We simply have to believe that they have the oil they say they have. This accounts for more than half of global proven reserves.

Since we are talking about oil – the primary mover and most important strategic resource in the world – we cannot afford to be so naïve. This is akin to the world economy buying a used car with the seller throwing a third of the brake parts away, but saying that they still work perfectly. Trusting him, we keep the pedal to the metal as we cruise along the highway.

Why would OPEC exaggerate?

It is easy to come up with reasons why the OPEC countries might withhold some of the information from their customers. These include:

- Local governments are totally dependent on high oil prices. Significant, market-driven moves to curtail oil consumption and increase alternative energy sources threaten these prices.
- Situations where current oil exporters admit that they have been lying for decades, would lead to unforeseeable and unpleasant consequences in global geopolitics and a total erosion of trust.
- Saudi Arabia would see its position as the trusted swing producer threatened.
- Governments would see their domestic credibility and powerbase erode, as those are largely based on oil revenues and trust in future oil revenues.
- There would be potential disruptions in the pecking order inside OPEC.
- It is not sensible to openly display any weakness in the volatile Middle East.

From the perspective of the oil producers, it is understandable that they might not be completely honest or open about the situation. According to game theory, a producer can only lose if it tells of its depleting reserves to others before it becomes clear to everyone anyway. The more interesting question is who profits from the continuing lack of healthy skepticism by the western nations? Or is it possible that the leaders of western nations simply are completely unaware of the risks, or unable to do anything about them?

Of course it is possible that the political reserves really do exist. But it is impossible to verify due to the surrounding secrecy. This uncertainty is the biggest problem in political oil reserves. The IEA builds its projections on top of these uncertain assumptions. Then our leaders base their decisions and policies, that reach far into the future, on IEA reports. If we base our future strategies largely on uncertain information, there is a high risk that we will make wrong choices. Perhaps we should instead base our strategies on the uncertainty of this information.

Paper oil of the oil companies

> I am becoming sick and tired of lying about the extent of our reserves issues and the downward revisions that need to be done because of far too optimistic bookings.
>
> Email from Shell's head of exploration to the chief executive (2004)

The international oil companies also have their share of paper reserves, as replacing produced reserves with new ones is one of the hardest tasks of a modern oil company. In 2004, the gas reserves of Royal Dutch Shell shrank by 20 percent (around 4 billion barrels of oil equivalent) with a simple statement from the corporation. The company had to pay hundreds of millions in damages to its owners who had bought the company stock (assuming too high reserves) between 1999 and 2004. In addition to its reserves, the company also lost its CEO.

The former CEO of Exxon decided to add the company's claims in the oil sands to its reserves, even though it was against the SEC (Securities and Exchange Commission) rules.[57]

Another subject entirely is how reserves today are added and backdated.

Backdating added reserves

What does it matter if the reserve additions in producing fields are recorded as oil found today, or as oil found on the year the field was discovered? The final amount of oil remains the same, and usually it is a simple case of moving 2P technical reserves into 1P proven reserves. This detail of accounting practices makes all the difference if we want to predict future oil discoveries and the ultimate amount of oil we will have, as these predictions are based on the trends of recent history.

To simplify a little, the trend of future oil discoveries seems quite different for the economists, depending on whether we have actually found new oil (which is false), or have simply moved technical reserves from already found oilfields into proven reserves (which is true). If we correctly backdate our reserve additions, we were peaking on oil reserves in 1980, and not now.[58] Nearly all officials and agencies book the reserve additions into the current moment, so it is no big wonder that the economists and governments see less reason to worry about oil discoveries. In reality, we found much more oil a few decades ago than we booked into reserves. Consequently, we have found much less oil in recent years than we have booked into reserves.

This is a key factor when someone says we have found much more oil than we have used in recent years. This is simply not true, as most of that oil was discovered decades ago in the form of the oilfield it is located in. To clarify, this is a discussion about peak reserves, not peak production.

Environmentalism, regulation, and mitigating climate change

A large proportion of the actual costs of using oil have been externalized, and are not paid at the gasoline pump. These costs include:

- the greenhouse gas emissions from the burning of fuels and their future costs due to worsening climate change;
- the pollution and particle emissions from burning fuels and the premature deaths and sicknesses they cause (millions annually);
- other indirect costs that increased traffic causes;
- the environmental damage of oil spills and other accidents and their economic costs.

Plentiful and cheap oil is plentiful and cheap largely because we do not include all of its true costs into its price. This is a market failure. It distorts the marketplace and creates incentives for people to use more oil than they otherwise would, as they do not incur the indirect costs of oil use. It is hard to prove that a child got asthma because of air pollution that resulted from increased traffic in the area, for example. It is also next to impossible to prove anthropogenic climate change (that is largely due to the burning of fossil fuels, among other things) as the reason behind an isolated extreme weather event.

In a capitalistic system like ours, it is often profitable to bend and break the rules. Indeed, in a competitive environment, whatever is possible will also eventually become necessary. Lead is extremely harmful to human beings even in small doses.[59] As a result it was banned in the U.S. in the early 1970s as an additive to gasoline. Still, some oil companies continue to manufacture and sell leaded gasoline in countries where it is still allowed.[60] The oil companies do not want people to cut their oil use and dependency as they would soon be in trouble with their huge, long-term investments. It is therefore logical for them to use vast sums of money to hide these hidden costs and to lobby against any attempts to try to include them in the price of oil. This has resulted in holes the size of a Hummer in the CAFE standard of the United States that have remained in the legislation for decades. One of the well-known holes is the possibility to register a vehicle as a light truck instead of registering it as a personal vehicle. This has made it possible to leave the SUV, or Sport Utility Vehicle, outside of the CAFE standard. There have been years of effort to close this gaping hole in the legislation. It is also a harsh truth that most of us want cheap gasoline. And even if we suffer from the negative costs, we don't want to think about them. According to one study, the correct price of gasoline should be around 3 euros per liter, or 15 dollars per gallon.[61]

On the other side of the fence, there are the environmental organizations and other NGOs (non-governmental organizations). They aim to include the externalized prices in the price of oil, which would make oil less attractive to use. They also try to limit and minimize environmental destruction by aiming to ban oil use and oil production in certain areas. All of these activities raise the price of oil and limit the amount of oil that is recoverable. For example 350.org, an organization focused on mitigating climate change had its part to play when in 2011 President Barack Obama sent the Keystone XL Pipeline back for environmental impact review. The purpose of the pipeline was to bring upgraded Canadian bitumen to the Gulf Coast refinery area. The fight goes on, and in his 2013 environmental speech President Obama was rather vague about Keystone XL, which left many environmentalists worried.[62]

If the production of bitumen or shale oil damages the environment, the locals have the right to try to prevent it. Some people even think that other life forms besides humans have some kind of right not to be destroyed at the altar of maximizing quarterly profits. Given that humanity is utterly dependent on other life forms and the services they provide us, this might not be a radical way of thinking. Furthermore, if the production of unconventional oil accelerates climate change even faster than the use of conventional oil, we all have a right to question the activity. The best way to question it is of course not to fill up at the gas station, and reduce our usage of fossil fuels. Oil will be produced only if someone buys it, and as long as enough people are buying, the job of the oil companies is to make it as plentiful and as cheap as possible. On the other hand, the global mechanisms of the oil market and its universal usability and properties make it a near certainty that the oil we refuse to buy will find another buyer somewhere else.

If environmentalism could be seen as making peak oil worse by limiting our ability to produce oil, there is also the other side to the story. Environmental protection as a broad concept helps to promote many activities that will reduce our oil use and dependency. These include the promotion of public transport, criticizing hyper-consumerism, promoting energy efficiency and conservation, and less energy-dense farming methods. Climate change mitigation often also helps to mitigate peak oil, even if the time span of these policies is usually far too long from a peak oil perspective. If we face peak oil largely unprepared, the changes in our societies and activities might be so fast and disruptive that the long-term climate mitigation plans and agreements will lose any relevance they might have had. Their remaining function will be to make people feel that the politicians and other decision makers are doing something about the climate problem, which will leave less for everyone else to worry about.

Commodities markets and their effects on oil prices

The oil trade, made up of a variety of market instruments, has its effect on the price of oil. It would be far too simple to claim that our high oil prices are due to speculators and institutional investors playing in the marketplace.

It would also be a stretch to claim that the price of oil is the result of supply and demand on an effective marketplace, without any regard for the players' meddling. This section brings some background to what happens at the marketplace, behind the scenes, and who is making it happen.[63]

Speculators

Speculation is essentially an activity where someone sells or buys something in the belief that it can be sold or bought back at a profit later. It is easy to blame the greedy speculators when the prices of raw materials, liquid fuels or other commodities rise and cause problems and civil unrest around the world.

What is often forgotten by these critics is that the speculators not only make money with rising prices, they also make money on falling prices. In other words, the job of these traders is to guess, or know, the direction that the price of a commodity will take. They seek the trends, acquire information from the markets and production around the world, and, based on that insight, they bet on either rising or falling prices for a given commodity in a given time span. In doing so, they give the rest of us valuable signals about whether we should increase or decrease the production or consumption of a given commodity. The speculators also bring liquidity to the marketplace. They are able to buy and sell even when there are not many other buyers or sellers. Speculators are an important part of the invisible hand that is said to move things along the marketplace. It is another question how well this invisible hand has been working lately, or how well it can be applied to commodities such as oil that have increasingly inelastic supply and demand. Speculators and Exchange-traded Funds (ETFs), are players in the oil market, but they are not able to produce, store, refine, resell or use the actual oil by themselves. What they trade in is *paper oil*. Paper oil in itself does not change any actual, real-world material flows, but it does change money flows and price levels.

Speculators earn their living with fluctuating prices, no matter whether they fluctuate up or down. Only a market with a stable, unchanging price of oil is disadvantageous to speculation. On the other hand, a marketplace where the price is not defined by any logical means is hazardous to speculators, since these kinds of fluctuations are impossible to predict. Of course the speculators will use loopholes or rock the boat if given the chance, as the rocking boat is where they thrive. Despite all this, it would be unfair to attribute rising oil prices simply to the greed of malicious speculators and dealers.

ETFs and institutional investors in the oil futures markets

A few decades ago, the main function of trading in oil was to buy or sell oil for refinement and consumption. When big players such as pension funds, investment banks, and other institutional investors entered the picture, the nature and the predictability of the oil trade were transformed almost completely. More changes were coming, as the 2000s brought oil to the electronic marketplace, making it available for trading to a vastly increased number of participants.

ETFs are funds that are traded in stock exchanges that normal people can relatively easily buy shares from. These funds do not want their paper oil delivered however. They only want to invest in oil, or be exposed to the price development of oil. In short, if they buy oil futures that mature in 3 months, they need to sell these futures before that time is due, and then buy new ones. This is called a "rollover." It is a similar event to when Greece needs to sell their debt again in the marketplace when their old debt becomes due and needs to be paid.

It is important to remember that investing in commodities in this way will only make a profit if the price of the commodity rises in a certain time period. Commodities such as oil do not pay dividends like many companies. Investing in commodities therefore relies on the assumption of a "bigger fool." This means that in order to profit, a bigger fool needs to be found to pay the higher price.

ETFs also function in a mechanical way. They buy futures according to their profile on a certain day no matter what the price is or how smart a certain trade may seem. This makes investing harder by distorting the marketplace. This happens when a big buyer (e.g. an ETF) enters the market at a certain moment with a commission to buy a certain large amount of oil futures, for example. This will be done with no regard to the price of those futures, and will in that sense be irrational from the marketplace's point of view. This can surely ruin the day for a speculator who did not see the event coming, and basically will make it impossible, or at least much riskier, for anyone to act rationally in the marketplace.

The need for funds to roll over their paper oil also creates an artificial pressure on the price of oil. This is further strengthened by the tightness (inelasticity) of the markets, as production is unable to grow to meet demand at a lower price. The rising price will lure more investors and funds into the bubble, which will eventually burst.

The actual buyers of oil that need their oil physically delivered have to pay the same inflated price as the investors do. This has created an interesting phenomenon in the commodities markets and the refining industry. Oil is by far the most popular energy investment. Gasoline, that is refined from oil, has been cheaper from time to time than the oil that it is refined from. This means that the refinery has had to pay more for the oil than it will get from the gasoline it will refine the oil into. Longer contracts help to mitigate the situation, but it is still problematic.

When we add the explosion of individual traders, many of whom are amateurs who do not necessarily operate rationally, into the marketplace, and the ease of trading due to the new electronic marketplaces, the situation gets really interesting. There is a lot more paper oil in the market than there is physical oil. The thought stretches the imagination for the layman, but it is perhaps indicative of the world of investing that nowadays has nothing to do with physical reality. Sadly, even though it is a world of its own, the events that happen in it also affect the real world.

As we are on the bumpy plateau of oil production, the spare capacity to produce oil tends to get smaller. As demand has proven to be inelastic to the price of oil as well, the price volatility increases. This is good for the speculator, and currently their activities often strengthen the volatility even further. As peak oil is realized and enters common knowledge for more investors, it will probably cause demand to stay high compared with supply. This will help to keep the prices high. On these grounds, it is easy to justify investing in these commodities and ETFs that give at least some kind of opportunity for profit from this otherwise unwanted phenomenon.

The price of oil

> You know there are no quick fixes to this problem. You know we can't just drill our way to lower gas prices . . . Yes, oil and gas, but also wind and solar and nuclear and biofuels, and more.
>
> President Barack Obama, February 2012

The price of oil is the result after all of these factors in its demand and supply have been accounted for. It is the easiest indicator of the oil situation: What is happening at the oil markets? Has the price gone up or down over a certain period, and what is the price now? On the other hand, given the short-sightedness of the markets, the price of oil tells us little about the longer trends.

The EIA commissioned a study in 2011 on the price volatility of oil.[64] It listed seven factors as the basis for the price of oil. They are, in random order:

1 the production trends of non-OPEC countries
2 the level of strategic reserves
3 the demand from non-OECD countries
4 the demand trends from mature OECD countries
5 local factors that create market disturbances, such as the weather, accidents or political upheaval
6 the trends of production from OPEC countries
7 financial activities in the market, such as speculation and funds.

Numbers 5 and 7 are short-term factors, while the rest affect the longer term. We have a few suggestions to add to the list:

1 The incomplete and unproven information the market has about oil reserves. The market will work only if it has enough correct information.

Currently, the market does not have a clue about the actual reserves of the OPEC countries.

2 The concealment of production getting harder. Markets do not get any information on the EOR methods that Saudi Arabia is using, for example.
3 Exaggeration of new oil reserves, mixing them with resources and the effectiveness of new production methods and technology. This is mainly done to attract investor money.
4 The risks, costs and limitations that environmental regulation and environmental accidents can have on oil companies' reputations.
5 With oil supply disruption, the price of oil starting to follow the most expensive barrel.
6 The weak correlation (weaker than previously thought) between oil prices and oil production, which leads to production not growing nearly as much as the rising price would suggest.
7 The effects that the oil price will have on GDP, unemployment rates and food prices for example, and the feedback loops these will cause to the demand for oil, which has probably been underestimated.

Does oil have a maximum price? Was the price of 147 dollars per barrel in 2008 the highest price that our economy could handle? In the longer run perhaps not, though there have been arguments that even the current price of around 110 US$ is too high. This would be worrying news, as the marginal price of new oil production was already 114 US$ back in 2012.[65] Robert Rapier, author of a book called *Power Plays*, takes an interesting approach in his article "Petroleum Demand in Developing Countries"[66]: "If you were given only three barrels of oil per year, how much would you pay for them?"

The average European consumes around 10 barrels of oil per year. If only a quarter of this was available, we would probably be ready to pay much more for this oil than we pay now. The citizens of many developing countries only consume around 2 barrels per year. It is likely that, given the chance to improve their living standards, they would be willing to pay much more for their third annual barrel than a European person would be for their eleventh. This possibility of relative economic growth will see to it that the long-term oil prices continue to go up. As expensive oil works as an effective plug for economic growth in developed countries, their growth possibilities lie more in their ability to rapidly make their oil use more effective and productive. The price of a new marginal barrel of oil production is perhaps around 90 US$. This would mean that the price would not drop much below that, because then it would become more profitable for the oil companies to buy oil from the existing marketplace to be refined instead of investing in new exploration and production.

Efficient and productive use of oil, and smaller dependency on oil will be the key competitive advantages that a society or a company can have in the coming years and decades. We need to take the maximum productive use out of every barrel of oil, and where oil is not absolutely necessary we should replace it as quickly as possible.

Notes

1 Hall, C., Balogh, S., Murphy, D. (2009), What is the minimum EROI that a sustainable society must have?, *Energies*, 2, 1, 25–47.
2 http://ourfiniteworld.com/2013/06/24/energy-products-return-on-investment-is-already-too-low.
3 Lambert, J., Hall, C., Balogh, S., Poisson, A., Gupta, A. (2012), *EROI of Global Energy Resources – Preliminary Status and Trends*, State University of New York, College of Environmental Science and Forestry, http://tinyurl.com/pbb6mqb.
4 Moerschbaecher, M., Day, J. (2011), Ultra-deepwater Gulf of Mexico oil and gas: energy return on financial investment and a preliminary assessment of energy return on energy investment, *Sustainability*, 3,10, 2009–2026.
5 www.westernresourceadvocates.org/land/oseroi.php.
6 www.energywatchgroup.org/fileadmin/global/pdf/EWG_Report_Coal_10-07-2007ms.pdf.
7 www.eia.gov/todayinenergy/detail.cfm?id=4030.
8 Vikman, R. (2012), *Shale Gas: The Promise and the Peril*, RTI Press.
9 Aucott, M. L., Melillo, J. M. (2013), A preliminary energy return on investment analysis of natural gas from the Marcellus shale. *Journal of Industrial Ecology*, 17, 668-679, doi: 10.1111/jiec.12040.
10 Weißbach, D., Ruprecht, G., Huke, A., Czerski, K., Gottlieb, S., Hussein, A. (2013), Energy intensities, EROIs (energy returned on invested), and energy payback times of electricity generating power plants, *Energy*, 52, 210–221.
11 EMROI comes from energy money returned on invested. It calculates the prices of the feedstock used in energy production, without human labor costs. It is often confused with EROI (or EROEI) in studies.
12 www.theoildrum.com/node/3910.
13 Pressurized water reactor (PWR). Another common reactor type is a boiling water reactor, or BWR.
14 Weißbach, D., Ruprecht, G., Huke, A., Czerski, K., Gottlieb, S., Hussein, A. (2013), Energy intensities, EROIs (energy returned on invested), and energy payback times of electricity generating power plants, *Energy*, 52, 210–221.
15 http://en.wikipedia.org/wiki/Integral_Fast_Reactor.
16 Morgan, T. (2013), *Life After Growth: How the Global Economy Really Works – and Why 200 Years of Growth Are Over*, Harriman House.
17 *Wall Street Journal*, 1 July 2013, http://tinyurl.com/nc24wfo.
18 http://aleklett.wordpress.com/2013/09/16/goldman-sachs-360-projects-to-change-the-world-english.
19 www.channel4.com/news/repsol-demands-10bn-compensation-from-argentina.
20 Perkins, J. (2004), *Confessions of an Economic Hitman*, Berrett-Koehler.
21 www.bakerinstitute.org/publications/NOC_IOCs_Jaffe-Soligo.pdf.
22 http://en.wikipedia.org/wiki/Oil_reserves_in_Venezuela.
23 www.usatoday.com/money/world/story/2012-05-28/venezuela-oil/55248628/1.
24 Gustafson, T. (2012), *Wheel of Fortune: The Battle for Oil and Power in Russia*, Belknap Press of Harvard University Press.
25 Aleklett, K. (2012), *Peeking at Peak Oil*, Springer.
26 http://moneytalks.net/topics/energy/7767-oil-astounding-drilling-projects-dollar-bills-for-a-dime.html.

27 Aleklett, K. (2012), *Peeking at Peak Oil*, Springer.
28 www.halliburton.com/ps/default.aspx?navid=2432&pageid=4983.
29 Aleklett, K. (2012), *Peeking at Peak Oil*, Springer.
30 Implications of worldwide population growth for U.S. security and overseas interests (The Kissinger Report), 1974, http://pdf.usaid.gov/pdf_docs/PCAAB500.pdf.
31 http://aleklett.wordpress.com/2012/09/12/will-saudi-arabia-become-oil-importer-by-2030-new-blogg.
32 www.bloomberg.com/news/2012-04-02/saudis-skip-arab-spring-as-nation-pours-money-into-jobs.html.
33 http://globalpublicsquare.blogs.cnn.com/2012/01/15/zakaria-why-oil-prices-will-stay-high.
34 http://physics.ucsd.edu/do-the-math/2011/10/the-energy-trap.
35 http://en.wikipedia.org/wiki/Exergy. Exergy has been popularized by R. Ayres and B. Warr in their papers and books on energy and its role in society and the economy.
36 *EIA Annual Energy Outlook* 2012 explains the relationship between Alaska's North Slope oil production and oil prices.
37 The break-even price reflects the minimum price that a producer must get in order to stay economically (and in this context, also socially) viable.
38 The artificially weak currency of China devalues its GDP. We have used the more relevant PPP-GDP (Purchasing Power Parity). The *World Factbook* (2012), CIA, www.cia.gov/library/publications/the-world-factbook.
39 Aleklett, K. (2012), *Peeking at Peak Oil*, Springer.
40 *World Energy Outlook* (2010) Key graphs (c). OECD/International Energy Agency 2010, p. 4.
41 Legget, J. (2005), *Half Gone*, Portobello Books Ltd, p. 150.
42 *The Oil Kings: How the U.S., Iran, and Saudi Arabia Changed the Balance of Power in the Middle East* by Andrew Scott Cooper (2011, Simon & Schuster) offers interesting insights into what happened behind the scenes at the top political level. Bankrupting nations in Europe, the global financial crisis, deepening recession, and the polarization of politics were familiar subjects then as now. The Shah of Iran was also well aware of the coming peak oil.
43 *New Scientist*, 28 June 2008, pp. 32–37.
44 www.energyrealities.org/detail/saudi-arabia-oil-reading-naimis-runes/erpF5E46AD452C5D0049.
45 http://rslissak.com/content/saudi-arabias-economic-needs-and-price-oil-joseph-mann.
46 http://extras.springer.com/2010/978-1-4419-5987-4/Chapter%2012.pdf.
47 Aleklett, K. (2012), *Peeking at Peak Oil*, Springer, p. 230.
48 www.telegraph.co.uk/news/worldnews/1532455/Iraq-would-be-terrorists-launch-pad-if-we-left-says-Bush.html.
49 www.guardian.co.uk/world/2007/sep/16/iraq.iraqtimeline.
50 www.reuters.com/article/2013/02/17/us-iraq-pipeline-attack-idUSBRE91G08I20130217.
51 http://gcn.com/articles/2012/04/23/iran-cyberattack-oil-industry-offline.aspx.
52 http://oilprice.com/Latest-Energy-News/World-News/Shamoon-Virus-that-Attacked-Saudi-Aramco-is-the-Most-Dangerous-to-Date.html.
53 http://gulfnews.com/news/region/iraq/iraq-s-baiji-oil-refinery-shut-by-bombing-1.767970.
54 www.eia.gov/countries/regions-topics.cfm?fips=WOTC&trk=p3.
55 www.vanityfair.com/online/daily/2012/03/lawrence-wilkerson-middle-east-nancy-schoenberger-israel-iran-nuclear-weapons.

56 Aleklett, K. (2012), *Peeking at Peak Oil*, Springer, p. 126.
57 Coll, S. (2012), *Private Empire – Exxon Mobil and American Power*, Penguin Press.
58 www.theoildrum.com/node/9389.
59 In addition to the many known harmful properties of lead, there has been an interesting discovery that indicates some level of causality between lead contamination and a general tendency towards violence in a society: www.guardian.co.uk/commentisfree/2013/jan/07/violent-crime-lead-poisoning-british-export.
60 Tamminen, T. (2008), *Lives Per Gallon: The True Cost of Our Oil Addiction*, Shearwater.
61 http://cironline.org/reports/price-gas-2447.
62 www.usnews.com/news/articles/2013/06/25/obama-says-no-to-keystone-xl-if-it-harms-environment.
63 More information about the oil market can be found for example in Dan Dicker's book *Oil's Endless Bid – Taming the Unreliable Price of Energy to Secure Our Economy* (2011, Wiley). It offers an interesting insider perspective to the oil market and their huge changes during the last few decades.
64 Ederington L., Fernando, C., Lee, T., Linn, S., May, A. (2011), *Factors Influencing Oil Prices: A Survey of the Current State of Knowledge in the Context of the 2007–08 Oil Price Volatility*, EIA.
65 www.forbes.com/sites/christopherhelman/2013/06/13/why-americas-shale-oil-boom-could-end-sooner-than-you-think.
66 www.theoildrum.com/node/9386.

3 What instead of oil?

Diminishing oil production

> The simple message is we can't develop the unconventional fossil fuels, tar sands being the first example, and have any hopes of stabilizing climate in the foreseeable future.
>
> Dr. James Hansen, NASA

The production of the currently producing oilfields is being depleted by around 5–7 percent annually. This means that we lose 4–5 mbpd of production every year. This production needs to be replaced with the following:

1 enhanced oil recovery methods
2 finding and developing new conventional oilfields
3 developing production from unconventional sources
4 growing and manufacturing biofuels and synthetic fuels.

EOR methods include pumping water, carbon dioxide or natural gas into a field to help increase its pressure. With these methods, the profit per expenditure factor gets progressively weaker; the field needs more and more water to get less and less actual oil from it. In the world of expensive oil, many of these methods become profitable, which leads to increasing recovery factors and faster production from old fields. New conventional fields are also being found and developed constantly. As a rule, however, these are much smaller and in more remote locations. The trend has been that a lot less new oil has been discovered during the last few years than has been used.

Iraq became the third largest oil exporter in 2012, after Saudi Arabia and Russia. According to a fresh report by the IEA, almost half of the growth in oil production in the coming decades has to come from Iraq. If we fail at this, the global oil markets are headed for troubled waters. To succeed, Iraq needs to double its current production to 6.1 mbpd by 2020. From there, the production needs to grow to 8.3 mbpd by 2035. This growth needs to happen throughout the entire infrastructure: transportation routes, drilling rigs, water transportation (around 8 mbpd), storing facilities, and professional personnel all need to increase. In addition there are the political problems and negotiations for contracts that would need to be solved immediately. Iraq also needs a society-wide update so that it can soak up the growing revenues from the oil as evenly as possible. It needs 70 GW of new electricity production capacity by 2035, as GDP is projected to grow fivefold. The 9 billion dollars invested in the energy sector need to almost triple to 25 billion a year to meet these goals.[1] The GDP of Iraq in 2011 was around 112 billion dollars.[2]

Unconventional oil includes oil sands, shale oil, and kerogen oil (oil shale). These are hydrocarbons that can be refined or upgraded to match conventional oil with reasonably little effort. Their production is very different from conventional oil production. For example they are often mined, not pumped.

Other substitutes for conventional oil can be synthesized from other hydrocarbons such as coal, natural gas, and biomass. The methods are one step further, and a couple of steps more expensive, than conventional oil production. The one used on a commercial scale today is Fischer–Tropsch synthesis (FT) that can turn virtually any hydrocarbon into liquid fuel, albeit expensively. More on FT is explained later in the book. The direct use of natural gas in transportation requires a significant modification of infrastructure and vehicle fleets.

We can also reduce our oil consumption by changing the power source of our transportation, with the obvious options being electricity and hydrogen. Neither of these is an energy source, but an energy carrier. This means that they have to be produced first from another energy source. That is to say, they are more about dealing with a liquid fuels crisis, and not so much about dealing with an energy crisis. A report commissioned by the U.S. Department of Energy (DoE), "Peaking of World Oil Production: Impacts, Mitigation, and Risk Management," commonly referred to as The Hirsch Report, was published in 2005. According to this, a society needs to start full-scale mitigation 10 to 20 years before the actual moment of peak oil in order to meaningfully soften the negative effects of depleting oil production. In 2011 Robert Hirsch, the author of the report, estimated that we would reach peak oil between 2012 and 2015.

Decreasing our oil use

As a result of the oil crises in 1974 and 1979, the developed world has reduced its oil use in the sectors where it was economically most profitable. Burning oil to create electricity has dropped to a small fraction of what it was in most western nations. This means that further reductions will be progressively more difficult to achieve. A focal point is to replace oil as a transportation fuel. More than half, 58 percent in 2004,[3] of global oil use happened in the transportation sector. It will be much harder to replace oil as a raw material for plastics and other synthetic fibers, in construction, and in agriculture.

Crude oil can also be replaced with another similar fuel. This includes bitumen from oil sands, kerogen from oil shale, biofuels such as biodiesel and ethanol, and synthetic fuels such as liquefied coal. According to Kjell Aleklett, these unconventional substitutes could account for 9 mbpd by 2035. They include:

- NGL (natural gas liquids and condensates)
- GtL (gas to liquids)
- CtL (coal to liquids)
- oil shale or kerogen oil
- shale oil
- extra heavy oil from Venezuela's Orinoco Belt
- bitumen from the oil sands in Canada.[4]

We could also switch the power source of our logistics, which currently is powered 95 percent by oil and internal combustion engines.[5] In urban areas, public transportation should be electrified and moved to rails where it is feasible, since these are much more efficient than personal vehicles. This also helps to reduce local air pollution. Of course, building rail into rural areas where they can be used much less is hardly commendable, and more efficient methods of moving people and goods around need to be innovated.

In reality, we will probably have to choose all of the above. We need to employ different substitutes, change power sources, make transportation more efficient and even reduce our transportation needs altogether. Unconventional oil sources will likely postpone peak oil by a few years and perhaps make the price increase less steep, but they are not likely to solve the peak oil problem. Resources are huge, and technology, together with high prices, has increased their proven reserves and production in recent years, but their production is still slow and expensive to grow. In addition, their EROEI is often weak, which makes them reasonable as a transitional fuel, but not as a replacement for cheap oil. Oil is also extremely valuable as a raw material. In principle, there is little sense in burning it just to move around, often unnecessarily.

What makes the situation harder is that we cannot afford to build multiple infrastructures on top of each other. Also, building infrastructure is always slow. The middle class, who often finance these kinds of transitions, may not see a reason for them, since they still have money to drive around in their regular cars.

Ultimately we will end up with one technology or another. There will be differences due to local factors such as available energy sources. No matter how much all of these solutions are advertised as the only way forward, not all of them will be implemented everywhere. In a world of depleting and more expensive oil production, it would be sensible to use our oil to reduce our dependency on oil; to build oil-free infrastructure and to move to a society with a diminishing amount of oil for it to use while we still have the chance.

Deepwater oil

Why are deepwater oil reserves presented as an oil substitute? Their quality is often close to conventional oil. Deepwater oil is included here for three reasons:

1 it is unconventional to produce;
2 it depicts our situation that is getting more and more desperate;
3 it is often presented as the solution to peak oil.

The definition of deepwater oil is quite flexible, as it ranges between 200–500 meters beneath the surface. The same is true for ultra deepwater oil, which is for example 1,524 meters beneath the surface. Often we have to drill thousands of meters beneath the sea floor before we get to the actual oil and gas. This is a job for the modern semi-submersible oil rigs such as the infamous Deepwater Horizon. They cost (in 2001 dollars) around a billion dollars apiece, and have operational costs of around 500,000 dollars per day. Technologically, they are among humankind's most complex and biggest mobile constructions, as is reflected in their price. The amount of energy and resources used to construct them is immense. Further, the complexity of a society and the technological ability needed to construct a semi-submersible is immense, and so is the amount of energy needed to keep it humming.[6] Perdito, a rig completed in 2008 in a Finnish shipyard that later headed for the Gulf of Mexico, was the height of the Eiffel Tower in total and weighed around 22,000 metric tons.[7]

Deepwater drilling has been described with the following analogy: It is like someone drilling from the roof of a two-story building, from a rig the size of a box of matches, with a tube the size of a hair, aiming at a spot the size of a dime on the bottom floor of the house. Of course, the top floor is filled with flowing water, and the bottom floor is filled with hard granite.

The majority of deepwater oilfields are concentrated within the so-called Golden Triangle. This includes the coasts of Brazil, West Africa (e.g. in Niger and Angola) and the Gulf of Mexico. In 2008 the average size of a deepwater field was around 150 million barrels (less than 2 days of global consumption). At the same time, the average discovery onshore was around 20 million barrels. Since then the average deepwater discovery has dropped to around 100 million barrels. Still, an increasing amount of our oil will be produced from deepwater and ultra deepwater. This will add to the risks of oil production.[8]

The Gulf of Mexico and the coast of Brazil

Like the North Sea, the Gulf of Mexico (GoM) is one of the most well-known waters for drilling oil. It has both deepwater and ultra deepwater oilfields and drilling activities. In 2009 the GoM produced around 1.5 mbpd for the United States. It is comparable to 2 percent of global demand or the amount that Libya produced in 2011 before its civil war. It has been estimated that around 10 Gb of oil can be produced from the deepwaters of the GoM. This amounts to about 4 months of current global demand. Around a third of new fields produce oil, while the rest produce mainly natural gas. A new test drilling operation costs between 50–200 million dollars, and around 70 percent of those will be dry holes that never produce anything. Doing business in the GoM, or any deepwater area, is both risky and expensive. Around 5 million barrels of oil leaked into the waters during the accident of Deepwater Horizon in 2010, killing eleven crewmen and causing damage to the marine environmental and coastal areas.

A few years ago a field named Tupi was discovered 400 km off the Brazilian coast. The field was later renamed Lula. Preliminary drillings have been done and production has started. Estimates on the reserves range from 6–8 Gb. Production is estimated to reach 1 mbpd in the 2020s.

The risks for environmental accidents are significant, and Brazil has a lot to lose, since the famous beaches of Rio de Janeiro are located relatively near. Subsalt fields discovered near São Paolo and Rio de Janeiro are estimated to have as much as 80 Gb of oil, which amounts to 2.5 years of global demand. Only a small part of these fields, like Lula, has been drilled and started producing. The coast of Brazil is one of the most challenging places to drill for oil that we know of. After 2 kilometers of water, there are 5 more kilometers of rock, sand and salt to drill through before oil is found. Petrobras, the national oil company of Brazil, is planning to invest 225 billion dollars between 2011 and 2015. Around 128 billion of this will be invested in deepwater exploration and production along the Brazilian coast. In addition to a number of drilling rigs, there is a need for 250,000 professional personnel by 2016. Brazil has strict limits on employment of foreign people and equipment in its projects, which can slow down the expansion of the projects.[9]

The risks of deepwater drilling off the coast of Brazil started to manifest in early 2012, when an oil leak was discovered on the Chevron-operated Frade field, one of Brazil's largest. In March, Chevron had to close the field down, which was producing around 60,000 bpd. Chevron now faces bans on drilling, massive fines, and even jail time for some of its executives for up to 30 years.[10]

Global reserves and production from deepwater

Jean Laherrère, an oil geologist, has estimated the ultimate recovery from deepwater to be around 150 Gb. The total global production from deepwater sources in 2010 was around 6.7 mbpd. This is estimated to grow by around 200,000–300,000 bpd annually for the next decade, and will replace the production of around 5–8 percent of the global depletion of conventional oilfields. Peak deepwater oil, at around 8–11.5 mbpd will be reached in the first half of 2020s, after which production will start to decrease.[11]

Arctic oil

> Why are we drilling in the Arctic to meet the growing demand in the future, when we should be weaning ourselves off oil dependency?
>
> Greenpeace on Shell's Arctic drilling in 2012

Arctic oil deserves a chapter of its own for the same reasons as deepwater oil. Even if it often matches conventional oil in quality, its production environment is something else altogether. There are a host of definitions for Arctic oil, but one of the common definitions is oil north of the 60° latitude. The contradiction between our worries over climate change and our thirst for more oil is plain for everyone to see. On the one hand more and more attention is being placed on the dire consequences of climate change, while on the other hand we are anxiously waiting for the Arctic ice to melt so that we can get to the supposed resources hidden beneath.

The Arctic is warming faster than any other place on earth. The IPCC's (Intergovernmental Panel on Climate Change) 2007 report estimated that the first summer without ice in the Arctic Ocean might happen within 20 to 40 years. Currently it seems likely that we will see that happening during the current decade. In 2012, the area of Arctic ice hit a record low. While this made news all over the world, it neglected the more stark fact that the volume of the ice has shrunk by as much as 75 percent in the past 2 decades. In addition to all this, the Arctic region has an increasing role in the game of global geopolitics, and the stakes are growing.

There are different estimates on the reserves. The most optimistic are from CERA, at above 100 Gb, while Colin Campbell and Jean Laherrère from ASPO estimate the reserves to be 50 Gb. The United States Geological Survey (USGS) estimated in 2008 that there were around 90 Gb of oil, 44 Gb of natural gas liquids and 1,670 tcf (trillion cubic feet) of natural gas. None of these have yet been proven with drilling, so they are more or less crude estimates.

Oil has been produced in the Arctic region since the 1970s, but in this book we concentrate on the new fields that are located in the Arctic Ocean. The methods of actual production in these areas are still somewhat vague, but it is painfully clear that the environmental conditions will probably be the hardest we have faced so far. Should we drill above the ice, or wait for the ice to melt? What kinds of risks do rogue icebergs present, and what is the best way to transport the oil to markets safely? BP recently cancelled its development project in the Alaskan Liberty field after 18 months as the budget kept growing. The aim was to have the field producing 40,000 bpd, and it was estimated to contain around 100 million barrels of oil. This was not the first attempt there, as there have been development projects on the Liberty field since 1998.[12]

Then there are the incredibly harsh climate conditions, and the seasonal melting and freezing of the ice to consider. How much risk are we willing to take to get just a little more oil? Accidents would be devastating, since responding to oil spills could prove to be next to impossible with no infrastructure in place.

There is also an additional problem of the oil industry's structure to consider. As the chain of contractors and subcontractors gets longer, it becomes more and more difficult to define who is responsible for what. The complex web of responsibilities slows down decision making in a crisis, as Tainter and Patzek present in their case of the Deepwater Horizon accident in their book *Drilling Down* (2011).

It is estimated that at least 100 billion euros will be invested in Arctic drilling by 2020. These projects will have a limited effect on the total oil production during this decade. Shell is one of the first ones to start test drilling, and the infrastructure to transport the oil does not yet exist. This means that commercial production may be at least a decade away. As Russia sees its current fields depleting by 0.5 mbpd annually, it will also have to drill more in the Arctic region to replace lost production.

The risks and costs of Arctic oil production offer a valuable window into mankind's thirst for oil, and how it is more important than almost anything else. We look to the Arctic in a desperate attempt to keep business-as-usual going for a little while longer, rather than to harness critical fuel to help us transition to a less oil-dependent society and infrastructure. Nobody even attempts to pretend that we would even try to lessen on our dependency on oil in the foreseeable future, since its consumption is estimated to keep growing.

Bitumen from the oil sands

> Ugliest stuff you ever saw . . . contaminated, non-homogenous and ill-defined . . . Bitumen is five percent sulphur, half a percent nitrogen and 1,000 parts per million heavy metals. Its viscosity [stickiness] is like tar on a cold day. That's ugly.
>
> Dr. Steven Kuzniki, Alberta University

What kind of image does *oil sand* bring to mind? Few of us have ever seen the oil sands, but all of us have seen the stuff that the "oil" in the oil sands is. It is called bitumen, and it is the main ingredient in products such as asphalt and roofing. Technically bitumen is very heavy (it can have more than 1,000 atoms in a molecule) and very toxic. It is the stuff left at the bottom of the oil barrel.

Some claim that oil sands are oil that has migrated upwards, closer to the earth's surface. Over the course of millions of years, bacteria have eaten all the lighter and easier to digest oil molecules, leaving us with the nasty, heavy stuff. Typically, around 10–15 percent of oil sand is bitumen.[13] It cannot be pumped, it cannot be used as a lubricant and it doesn't burn easily, given that numerous forest fires have failed to ignite it.[14]

Bitumen is too thick to move through pipelines. Therefore it first needs to be upgraded. This is often done by removing the heavier coke and adding hydrogen. The energy and hydrogen needed usually come from natural gas, and the process has been likened to turning gold into lead. Bitumen can also be made more liquid by mixing it with synthetic oil that has already been upgraded, or by diluting it with lighter condensates. For example, the oil in Venezuela's Orinoco Belt was recently redefined from bitumen to extra-heavy oil and added to the nations' oil reserves. This made Venezuela the number one in global oil reserves. Canada also counts their bitumen (around 175 Gb) as oil, and therefore it has third biggest reserves, with Saudi Arabia falling between the two. The rise in oil prices and improvements in technology have caused some scrambling of the top five countries in terms of oil reserves.

Resources and production of oil sands

The largest oil sand deposits are in Alberta, Canada. There are smaller deposits elsewhere, for example in Russia. More than 80 percent of recoverable oil sand reserves are estimated to be in Canada, and it also has most of the current production. The total resources of Canadian oil sands are estimated at around 1,700–2,500 Gb. Of this, around 175–315 Gb are recoverable reserves. The global reserve estimates are 220–395 Gb, enough for around 8 to 13 years with current consumption rates. In reality, as the production capacity will likely stay below 2 Gb per year, there is enough bitumen in the oil sands to last for a century or more.

Bitumen accounts for more than 40 percent of Canadian oil production. As the conventional fields are depleted and oil sands production expands, this share will likely rise in the coming years. Alberta's oil sandbox is the biggest energy project in the world. It also offers one of the best terms for the oil companies to operate, for example in terms of royalties. Nikiforuk's book *Tar Sands* (2010), and many similar articles published on the topic, remind us of a boomtown during a gold rush. The horses have been replaced with SUVs, the whiskey supplemented with harder drugs, and the small towns are replaced with a massive real estate pricing bubble. Of course, the picture shifts depending on who you ask.

Production from the oil sands was around 1.9 mbpd in 2012. The production is estimated to grow substantially, but it is not as easy as drawing a growing line on a chart. The needed additional infrastructure spans the whole sector from trucks, water pipelines, upgraders, and transportation infrastructure to housing, schools and other services needed by the workers and their families. CERA estimates that by 2025 the oil sands will produce between 2–4 mbpd, depending on different developments and prices.

Most of the bitumen is produced by two alternative methods. One is surface mining (51 percent in 2011) used for deposits up to 75 meters deep.[15] In this method, first the top soil is removed and stored for later use. After that, the oil sand is mined and transported to a station where it is heated with water. After this the water, oil and sand are separated and collected.[16] The water is recycled as much as possible, the sand is stored, and the bitumen is transported to be upgraded. Around 75–85 percent of the bitumen in the sand can be recovered, which is a much higher percentage than conventional oilfields.[17] [18] Around 20 percent of recoverable reserves can be produced by mining, so the share of bitumen produced from oil sands by mining will get smaller in the future.

The Caterpillar 797 trucks that are instrumental to the mining operations are impressive. At around 5 million dollars apiece, you get 4,000 horsepower with which to move 363-ton loads of cargo at 40 mph. The F-series has a 20-cylinder, 4 times turbocharged Cat C175–20 ACERT diesel engine. The 59/80R63 XDR tires, developed together with Michelin, are the largest in the world and cost around 42,500 dollars each. The trucks are the height of a small apartment building; so big that they must be assembled on site. Their fuel tank holds 3,785 liters of fuel.[19] Compared with the 797, which burns 7,800 liters per 100 km[20] (0.03 mpg), a Hummer seems like the green car of the year.

The other *in situ* method of producing bitumen from the sands (49 percent of production) is called Steam Assisted Gravity Drainage, or SAGD. In this method, a hole around 300 meters deep is first drilled, and then horizontal pipes are laid, each one around 1 kilometer long. Next, hot (200°C) pressurized water and steam are pumped into the pipes. The water soaks into the surrounding sand, heating up the bitumen. This makes the bitumen more liquid, and it drains into the collecting pipes placed below. Around 60 percent of the bitumen can be recovered with SAGD, and the damage to the above-ground environment is likely to be less than with open pit mining. SAGD is however the more energy-intensive of the methods of bitumen extraction from oil sands.

Expanding production of the oil sands is only profitable if WTI (West Texas Intermediate) oil prices are 70–80 dollars per barrel. CERA says in its report that over 70 percent of the new projects were frozen when the oil prices suddenly crashed to 40 US$ during the 2008 economic crash. The cost of adding a further 1 mbpd of capacity is 126 billion dollars according to CERA. This calls for a lot of trust in constantly high prices of oil from both oil companies and investors. It also requires economic growth that might not happen with such high oil prices.

The price of current production is also affected largely by the prices of natural gas, since it is widely used to produce heat, provide hydrogen, and is burned for electricity. Some estimate that the SAGD process will grow to account for 40 percent of Canadian natural gas usage by 2035. This demands huge investments in gas pipelines and puts pressure on gas production in Canada.

Costs of production are also externalized to the local environment and tax payers. According to the book *Tar Sands*, Canadians collected a total of 150 million dollars in royalties between 1966 and 2002 and used 4 billion dollars to clean up the mess from abandoned mines. The funds collected to counter bankruptcies seem rather small, at only 10 percent compared with the actual costs. At least the royalties seem to be increasing. In the fiscal year of 2011/2012, the government of Alberta collected about 4.5 billion dollars in royalties from the oil sands.[21]

The EROEI of bitumen has been estimated to be around 3:1 (from bitumen to gasoline) or 6:1 (from bitumen to crude oil). These estimates probably don't include the energy needed for remediation efforts. What does this kind of EROEI mean? It means that we cannot run anything resembling our current industrialized society with bitumen. An EROEI of 3:1 means that a third of the energy has to be used in the production of energy. Bitumen's value as a liquid fuel is higher than as an energy source, which makes its production easier to justify. On the other hand, the natural gas used as the main energy source is also valuable and has a host of other possible uses.

Environmental footprint

The local environmental effects of open pit mining are immense. Some land reclamation projects have been made, and even been certified as successes. But will these reclamations fall to tax payers if the oil producers go bankrupt? The tailings ponds are another glaring problem. They are used to store the water that could not be recycled in the production process. This toxic sludge is placed in these ponds to settle for several decades. There are plans to require that tailings are reclaimed at the same rate, or faster, than new ones are produced.[22] Environmentalists estimate that in 2007 more than 11 million liters of waste water from tailings ponds leaked into the surrounding environment each day. This was set to grow many times over by 2012.[23]

According to industry, the local environmental damage of SAGD production is much smaller. This view has been contested, because SAGD needs a much larger and more dispersed infrastructure above ground. Both methods aim to recycle the water used, but approximately 2–4 barrels of fresh water is used for each barrel of bitumen. This water is mainly drawn from the nearby Athabasca River. The estimated consequences of this use of water vary greatly. Some say that the current usage cannot be increased without serious damage to ecosystems and people living downstream. Other reports say that the usage is only a few percent of the river's flow, even in the driest season, and even if the usage is increased according to plans.[24]

Greenhouse gas emissions from the use of bitumen also vary according to the source. They are estimated to be 1.5–3 times larger than emissions coming from conventional oil. This will make our climate impact even uglier if we start to rely more on bitumen instead of conventional oil. Other pollutants, due to upgrading and refining, are 2–3 times higher compared with conventional oil. In bitumen production areas, the local air can be worse than the air in China's big cities. The best thing about it, from the western nations' point of view, is that the bitumen is located in Canada, which is a western democracy.

The estimates of future production growth suggest that bitumen could replace 1–4 percent of the global depletion of conventional fields during the next decade. We have already gone so low in our global and faithful oil addiction that we are forced to dig dirty sand for our next fix. The image of a junkie going through last night's trash pile in the hope of a soothing relief is not too far off.

Light tight oil or shale oil

Shale oil, or more accurately light tight oil (LTO), is a good example of how technology and high prices bring new oil to the markets. We have been aware of this oil for a long time, but its production has not been profitable or even possible before now. It is often called shale oil in the media, which is a name that has also been used for kerogen that we will discuss later in the book. This mix-up is unfortunate, since these two are quite different, especially since their methods of production are totally different, as are the lead times of production. For example, the news stories about the recent discoveries in Israel, estimated to be between 150–250 Gb, have left many with the impression that this is "shale oil" (meaning LTO) when in fact the discovered "oil" was kerogen.[25]

LTO is, as its name suggests, a collection of light hydrocarbons. In addition to shale, tight oil can also be produced from sandstones and carbonates. It is often produced together with shale gas with horizontal drilling and hydraulic fragmentation, which is a process called fracking or hydro-fracking. Depending on the field, more or less of the hydrocarbons come out as methane (shale gas) or other, heavier, hydrocarbons such as ethane, propane, butane and so forth. These heavier molecules make up the wet gas and the oil part of the production.

Tight oil resources and production

According to a recent EIA study[26] there is around 58 Gb of technically recoverable unproven resources in the United States, and 345 Gb for the world in total. This estimate still excludes potentially large formations in the Middle East and Central Africa. The devil, that the mainstream media tended to overlook, was in the detail, including the following facts:

- This was an initial assessment.
- The assessment was done using production data from U.S. shale formations, and extrapolated from there to the global scale. This makes the data shaky, since shale formations vary greatly in quality. Indeed, the EIA warns that the estimates it gives are uncertain. According to the EIA, between 1 and 10 percent (usually 3–7 percent) of total resources can be produced.
- These are mostly unproven resources, not proven reserves. The difference between these two can be substantial, but the picture people get from reading the news often is that they are the same.
- These resources are estimated to be technically recoverable, which means that the price of extraction is unknown (not surprisingly, since most of these resources have not even seen any preliminary drillings yet).
- The time span until these resources and formations can see actual production is uncertain. The reasons can be found both above ground and underground. In most countries, the mineral rights belong to the state rather than the property owner (as is the case in the U.S.). There is uncertainty on the environmental effects of fracking, which has been banned in some countries such as France. The geologies of the formations differ wildly, which can make production slow and expensive, as we have seen in Poland.

More than 80 percent of shale oil production in the U.S. is done in two areas: the Bakken formation in North Dakota (Williston basin) and the Eagle Ford formation in Texas. According to a recent USGS study,[27] Bakken (together with the Three Forks formation, also in North Dakota) holds 4.4 Gb of reserves (95 percent probable) and 7.35 Gb of technically recoverable resources (50 percent probable). Eagle Ford has previously been estimated by the EIA to yield 2.46 Gb of oil. California is estimated also to hold around 15.4 Gb (64 percent of the total U.S. shale oil reserves), but there is not yet significant production. The geology is very complex, and the local environmentalists are a powerful lobby.[28] In the spring of 2014, the EIA slashed its reserve estimates for Monterey shale in California from 13.7 Gb to 600 Mb, representing a huge 96 percent drop in reserves.[29] This could have dire consequences not just for the U.S. and its plans for energy independence, but for the shale energy reserves worldwide, since they were estimated largely by relying on the production and recoverability of shale energy in the United States.

Compared with many industry estimates, these numbers are very conservative. Others are even more cautious. David Hughes estimates that Bakken will yield around 2.8 Gb, and Eagle Ford another 2.23 Gb. Further, he argues that shale oil is not common among shale formations, since the majority of it is produced from just two plays, with the rest of the plays being marginal.[30]

The production of shale oil has grown rapidly in the United States:

- 2010: 380,000 bpd
- 2011: 620,000 bpd
- 2012: 1,200,000 bpd[31]
- 2013: ~2,700,000 bpd.[32]

This growth is likely to be slowing down. According to David Hughes' report, "Drill, Baby, Drill," in which he studies the actual production data from thousands of fields, North American shale oil from the current plays can grow to about 2.3 mbpd, and will peak there between 2016 and 2018. After that it will start to decline, as potential drilling locations run out. Other estimates range from 1.8–3 mbpd in the next few years.[33] This growth will offset maybe 4–7 percent of global depletion in current fields in the coming years. In the longer run, a lot depends on the global scale of production ramp-up.

As we can see, the actual growth of tight oil production has surpassed many of the predictions made just a few years ago, and done it much faster than predicted. Why, and what does this meas? The capex (capital expenditure) investments of oil companies have grown more than fivefold since 2000 and a substantial part of those investments have gone to tight oil and gas. In this sense, the amount of investments creates an ominous shadow over the impressive production gains. It becomes even more ominous, since the current oil prices have not been enough to pay for those investments. Many of the big oil companies have announced that they are not growing, but cutting their investments for 2014. Shell is cutting its investments by almost 20 percent, ExxonMobil by roughly 10 percent, Chevron by around 5 percent and Eni by 5 percent.[34] Most analysts and models have expected the investments to keep growing, so it will be interesting to see what this will do to oil production in the longer run. If investments are cut in tight oil production, the production levels could start to level off and even decline quite soon.

The recent crash of natural gas prices in the U.S. has played a big role in the increase of shale oil production. As the associated liquids that are produced with shale gas started to play a bigger role in the profitability of a well, and as the operators found that they could use the same methods to drill for shale oil as they used for shale gas drilling, a lot of the drilling rigs were moved to more wet areas in Bakken and Eagle Ford. As previously drilled but uncompleted gas wells have now largely been completed, the shale gas glut has started to disappear, and the prices of natural gas are again rising. At some point, as the currently operating shale gas wells are being depleted, there may be another deliverability crisis. Bill Powers writes in his book *Cold, Hungry and in the Dark* (2013) that this might be just a few years away. If not that, then at least the rising natural gas prices will take drilling rigs and resources away from shale oil, which will in turn affect its production growth.

The production profile from a shale oil (or shale gas) well is quite different from a conventional oil well. It rises fast to its initial production (IP), which is usually also its maximum production. After that, the well starts a rapid depletion. In 2 years the production will have fallen by 80–90 percent on average. After 5 years, a well in Bakken is producing 33 bpd on average. Taking all the producing wells together, they deplete at an annual rate of about 40 percent, compared with the 5–7 percent of conventional wells. Also, as the plays mature, the average IP starts to decline as the best spots are drilled first. This leads to faster and faster drilling just to keep production at current levels, let alone grow. In 2012, 25,000 new oil wells were drilled in the U.S., while the same level of production took only 5,000 new wells in 2000.[35] A well costs around 8–10 million dollars to drill in Bakken. To keep production at current levels, around 8 billion needs to be invested every year. In Eagle Ford, the depletion is even faster.[36] Even a short pause in drilling, as was experienced in Bakken in December 2010 due to bad weather, leads to a production drop.

Global resources

Outside the United States, there has been much less serious exploration and production of shale oil. As mentioned previously, the EIA has an initial estimate of around 287 Gb of technically recoverable shale oil resources outside the U.S. The top countries are:

1	Russia	75 Gb
2	United States	58 Gb
3	China	32 Gb
4	Argentina	27 Gb
5	Libya	26 Gb
6	Venezuela	13 Gb
7	Mexico	13 Gb
8	Pakistan	9 Gb
9	Canada	9 Gb
10	Indonesia	8 Gb.

In Europe, there are estimates that the shale near Paris also might have a few billion barrels. None of these resources outside the United States have been proven and almost none have been drilled. Yet often the mainstream media fails to do proper background checking on what it reports. A good example of this is the recent coverage of West Siberia's Bazhenov shale. The reported estimates on reserves and resources vary by orders of magnitude:

- The Russian oil giant Rosneft has estimated 22 Gb of recoverable reserves.[37]
- Another estimate places recoverable reserves at 365 Gb.[38]
- USGS estimates the total supply at 80–140 Gb.[39]
- Bank of America Merrill Lynch gives a wild preliminary estimate at 60–140 Gb of total resources.[40]
- Wood Mackenzie reports total resources (oil originally in place) to be 2,000 Gb, which is 5 times the amount Bakken has comparably.[41]
- Some[42] go even further by estimating first that there are 126,000 Gb of oil equivalent (BOE) in biomass at the Bazhenov formation, and the area is 80 times the size of Bakken. From that they extrapolate that the recoverable reserves could be around 1,920 Gb, which is more than the current known global reserves and almost 100 times more than Rosneft's estimate.

ExxonMobil has agreed that it will start developing Bazhenov together with Rosneft in the near future. The price of production is estimated to be 30–80 dollars per barrel. Oswald Clint, an analyst, says that if we got 300 drilling rigs to Bazhenov, it might produce 1 mbpd by 2020.[43] Clint does not tell us where the rigs would be produced. Those 300 rigs amount to a substantial amount of hardware, at almost 10 percent of the global fleet of 3,600 rigs. In 2013 two-thirds of this fleet were producing in North America, of which half were fracking for shale.[44] As shale oil wells deplete very fast, the transfer of rigs from one location to the next would mean that the production in the first would start crashing. Of course, we can manufacture more drilling rigs, but we would also need well-trained personnel, and there are limits to how fast we could add to the fleet. Despite all this, Bazhenov shale could produce significant amounts of oil and gas for decades. The geopolitical implications are immense, as are the effects on climate change if we ended up burning all of these hydrocarbons.

In early 2013 Rosneft and others were still waiting for tax reliefs on shale oil exploration from the Russian government.[45] These reliefs are quite likely to arrive, as the Russian government announced tax reliefs on deepwater drilling by 2014. A government spokesperson also predicted that shale oil would get similar tax reliefs at around the same time.[46] The current prices of over 100 US$ per barrel seem too low for shale oil production to be profitable enough, if the oil companies are to also pay taxes and royalties for the oil they produce. Big international oil companies have been cautious, since Russia's reputation as a stable place to do business is poor. Historically, the foreign partners have faced frustration by constantly shifting bureaucracy and politics with suddenly worsening contracts and terms.[47]

The production and the break-even price of shale oil both face controversy with the views of many shale optimists. In early 2013, PriceWaterhouse Coopers (PwC) published its report[48] on future oil production and prices that seemed to forget basic mechanisms of the marketplace. According to the report, shale oil could grow to 14 mbpd by 2035 while the price of oil would drop significantly from what it is now. How is this possible, if the most promising fields such as the Bazhenov require tax reliefs, even at current prices, to be developed? PwC fails to mention how, but it seems that increasing costs of production would need to be externalized for someone else to pay for.

The hype around shale oil and talk about the energy independence of the U.S. grew as the presidential elections of 2012 drew closer. This talk might prove to be hollow. In spite of billions in investment and a growing rate of drilling, the Bakken formation in Montana has seen a decrease in production for several years. From its peak in 2006 at 100,000 bpd the production has fallen to around 60,000 bpd. IP of new wells has decreased to half of what it was in 2002,[49] so it gets harder and harder to replenish depleting production from older wells.

So far, the biggest profits in the shale business seem to be in the real estate and drilling rights business.This will mean bursting price bubbles. Production of shale gas has not been profitable recently. As Rex Tillerson, the CEO of ExxonMobil said: "We are all losing our shirts today. You know, we're making no money. It's all in the red."[50] As the promise of shale gas has been eroded with falling prices and fast well depletion, so have the assets of many companies been eroded. Here are a few examples of recent losses, from Bill Powers' 2013 book *Cold, Hungry and in the Dark:*[51]

- In the fourth quarter of 2010, Range Resources reported a decrease in the value of its Barnett Shale assets by $463.2 million prior to their sale.
- In the second quarter of 2012, BHP Billiton reported losses in its Fayetteville Shale assets (which it purchased from Chesapeake Energy in February 2011 for $ 4.75 billion) of $2.84 billion.
- Exco Resources (NYSE:XCO), a very active driller in the Haynesville and the Marcellus, reported a loss of $704 million in oil and gas assets in the first half of 2012.
- In the second quarter of 2012, Chesapeake Energy reported a loss of 4.6 tcf from mostly proven but undeveloped reserves in the Barnett and Haynesville shales.

Lately there have been more and more signs of the bubble bursting. According to Bloomberg, "North American oil and gas deals, including shale assets, plunged 52% to $26 billion in the first 6 months from $54 billion in the year-ago period."[52] Meanwhile Shell reported a loss of $2 billion US$ worth of its assets and put its acreage in Eagle Ford shale up for sale as gas prices fell and the acreage turned out to have less potential than originally thought. Shell CEO Peter Vosel (retires in 2014) has called the company's 24 billion dollar investments in unconventional oil projects in North America a disappointment.[53] The total investments in shale have been in excess of 500 billion dollars annually, and a lot of these investments were financed with borrowed money.[54] The winners have been the investment banks on Wall Street, as they have had a field day with large transactions and corporate restructurings, from which they gained significant fees.[55]

The oil produced from shale is light and of high quality, and therefore it gets a price premium. It has been estimated that oil prices need to be around 80 US$ per barrel for production to be profitable[56] (including all expenses and profit for investment). The IEA has estimated that production costs alone are around 50 US$ per barrel.

The light condensates produced have found an eager customer in the Alberta oil sandbox. The condensates are used for diluting the heavy bitumen to get it moving through the pipelines. If the production of oil sands is raised to 3.5 mbpd, the daily requirement for condensates would climb to around 0.5 mbpd.[57]

Environmental impacts of shale oil production

Shale oil is light and easier to refine than heavy oils or bitumen. This would indicate that its greenhouse gas emissions would be comparable to conventional oil. Since a significant part of the production is natural gas, there will be methane leaks. There is an ongoing debate on how much of the gas leaks, with the higher estimates being between 4 and 8 percent. Even if the total leaks from production, transportation, storage and end use amounted to just 1 or 2 percent, it would mean that its emissions would be comparable to coal use. This is because methane is a much more potent greenhouse gas than carbon dioxide. If there is no nearby market for the associated gas, as in Bakken, it is flared. While natural gas only has around half the CO_2 intensity compared with coal per energy unit, flaring does not offer any benefit to society. Leaks from production and transport can be minimized, but this requires tough regulation and enforcement, and the use of best practices.

The local environmental damage happens both underground and above ground. Below ground, the biggest worries have to do with the injection of the fracking fluid that contains hundreds of chemicals. While there is not yet much evidence of these chemicals migrating to aquifers, it is a powerful image that environmental groups use. Often the shale production happens hundreds of meters deeper than where the aquifers are located, so the risks should not be great. The more immediate harm and risks might happen above ground. With tens of thousands of cubic meters of chemical-laden fracking fluid used in each production well, transportation poses a problem, as water needs to be hauled to the site. Much of it is collected and recycled, some is deposited in deep wells drilled for that purpose, but a substantial part needs to be transported to a wastewater treatment plant. Often these plants are not properly equipped to handle the chemical cocktails of fracking water, and traffic accidents can happen. The drastic increase in truck traffic is maybe one of the largest drawbacks. Roads need repairing much more frequently, the amount of diesel and its particle emissions make people sick, as does the dust and constant noise. It has been estimated that the royalties the communities get from the shale oil and gas production cannot even cover the added expenses in many places, let alone leave something for a rainy day.

The industry is not sitting still. There are developments of less harmful or even harmless fracking fluids. Propane has also been used instead of fluids, which enables collection of the propane afterwards. Increased knowledge from studies, and the resulting regulation, will play a crucial role in both the volume and price of future production and its environmental footprint. The price of production is the defining factor on the minimum market price of gas, and therefore also the amount of reserves that can be economically produced.

One further question is the earthquakes that have been traced back to fracking. They have been weak in many places but what will be the situation in the unstable Californian region that is said to hold two-thirds of the technical U.S. shale oil reserves?

Kerogen from oil shale

Have you ever tried to squeeze oil from a rock? Kerogen is organic matter that can be produced from oil shale. There is an immense amount of this oil shale, and the kerogen it contains around the world. The most significant production of oil shale has been done in Estonia, where it has been burned in power plants like bad quality lignite to produce heat and electricity. Israel is also planning to dig into its kerogen resources, with planned production of 50,000–100,000 bpd of production by 2020.[58]

From a geological point of view, kerogen is crude oil that has been left uncooked. It has never ended up in the right geological conditions in order to develop into liquid oil. The molecules in kerogen are often more than 1,000 carbon atoms long (gasoline has around ten carbon atoms). These molecules are heavy and hard to break down. Kerogen in itself is a nightmare to refine. It needs to undergo the same kind of heat and pressure, only accelerated, that nature would have applied over the course of millions of years in the right kind of geological conditions.

In public oil and energy discussions, this oil shale is often mixed up with shale oil (light tight oil) since both of them are located in some kind of shale. Sometimes this is due to ignorance, and sometimes it feels like these mistakes are made on purpose. These two are completely different things. Drilling for LTO is totally different from mining oil shale. Also, the massive kerogen resources are often confused with the good quality and reasonably efficient production of LTO. This adds up, in people's minds, to a massive and easily producible shale oil reserve, that in reality does not exist. To make it clear, we refer to it as kerogen in this book.

The current (if largely unproven) estimates are that there are vast quantities of kerogen around. There are around 600 known deposits around the world. The USGS updated its estimates on total kerogen resources for the U.S. in a recent report.[59] The report claims that there are 4,285 billion barrels of oil originally in place in the Green River formation. This is more than previous estimates of global kerogen resources of around 3,000 billion barrels. The potential reserves that end up being produced depend greatly on what quality of shale can be produced economically. If only the high grade shale that has more than 25 gallons of oil per ton of shale can be produced, those reserves would add up to around 350 billion barrels. If a lower quality of shale with 15 gallons per ton is produced, the amount climbs to more than 1,100 billion barrels, which is almost the same as current proven reserves. The report adds that it is as yet unclear how this kerogen can be produced, or if it will ever be produced. There are estimates that kerogen production could grow to 1–3 mbpd in the coming decades if water use, electricity and environmental considerations can be resolved. Earlier estimates of global reserves were significantly smaller, at around 160 Gb.[60] The estimates between total resources and proven reserves that can be produced at current prices and technology vary greatly for reasons such as:

- Different countries use different rules for reporting, and there are few independent estimates.
- The kerogen content of different oil shale deposits varies between 5 and 40 percent, which greatly affects the economics of production.[61]
- The production methods are still in development.
- The economics also depend on oil and natural gas prices.

There have been attempts to get kerogen production going since World War II. The biggest obstacles have been economics, both in energy and in currency. Cooking and separating kerogen from oil shale is nowhere near as easy as getting oil under its own pressure from a conventional well. There is almost no commercial production of kerogen oil in the world, so it is unlikely to offer much relief for peak oil in this decade.[62]

The traditional method of production goes more or less as follows. First, the oil shale is mined. Then it is crushed, after which it is heated up to a few hundred degrees Celsius (the minimum is 300°C with the optimum between 480°C and 520°C). This will turn the kerogen to a liquid, after which it is collected. The remaining waste rock can be used for making concrete or bricks, but with this comes the problem of logistics. The amount of waste is huge, and transporting it long distances is economically unviable and demands a lot of energy. The rock also has a kind of popcorn effect, as it takes 20–30 percent more space than the original oil shale, so we cannot simply put all of it back where it came from.

The more modern method, still under development by Shell, for example, is an *in situ* process. In it, the oil shale is first fractured, and then heated for several years underground. After heating up, the kerogen starts to flow, and it is collected in the pipelines built underneath. This kind of method is planned for Israel's future production.[63]

As oil prices shot up during the 1970s' oil crises and became a more substantial part of U.S. GDP, the production of kerogen started to gain more interest. Billions of both public and private money were poured into the research and development of kerogen production. In the end, the economic and technological hurdles were deemed too high, and the projects were buried. The subject keeps resurfacing every time oil prices rise high enough. With peak oil at hand, production of kerogen has been dug up once again as a possible remedy. It is telling that over 100 years ago a newspaper in Colorado had the following headline, "Oil Shale Development Imminent."

The break-even price of crude oil that would make kerogen production profitable varies greatly, depending on whom you ask. The EIA has estimated it to be 35–54 US$ per barrel, with the IEA raising the bar to around 60 dollars, while the RAND Corporation places the price at 75–90 US$ (these are all in 2005 dollars; to get an approximate 2012 value, add 15–20 percent). This break-even price has historically risen over time, so that it is always a little higher than the current price of oil. One of the factors behind this is that when the prices of oil and natural gas rise, they also increase the costs of producing many unconventional oil resources. This makes the investments a moving target, and with high upfront costs, investments in kerogen production look expensive and risky.

The amount of net energy obtained from its production is the final arbiter of how sane the production of kerogen is from society's point of view. Estimates of the EROEI are between 0.7 and 13:1, with 2:1 as an example.[64] The best case gives kerogen a decent EROEI, but most likely these are extreme cases that cannot be replicated widely. It is often cheaper and easier to manufacture oil by liquefying coal than cooking it from oil shale.

There are ways to make the traditional method of production more efficient. We could use waste heat from other processes to preheat the kerogen, but this requires that a suitable coal plant or nuclear plant (for example) is located nearby. There is also the possibility to use the waste heat from the process to produce electricity. These kinds of combined heat and power plants, with some use found for the waste rock, could make kerogen somewhat reasonable and profitable to produce in a world of expensive oil.[65]

The production, transportation, and use of kerogen as a fuel emit more greenhouse gases than the use of oil, coal or natural gas. When burned for heat, the emissions are comparable to bad quality lignite, which is the worst kind of coal there is. The total emissions are greatly affected by the production method used for the energy used in the production process.

Depending on the method of production, there are other pollutants, such as sulfur dioxide, that may be released. The production also uses large amounts of fresh water, with estimates ranging from 1–3 barrels of water used per barrel of oil produced. Wastewater from the processes amounts to 7.6–38 liters for every ton of oil shale processed.[66] If there is no use found for the processed rock, it will remain as waste at the site.

In addition to the different unconventional oils presented above, it is also possible to synthesize oil from nearly any organic matter or hydrocarbon. There are many processes: fermenting alcohol, extracting bio-oil from oil plants, gasifying solids such as biomass or coal, and then liquefying them. Next we look at the most common method of liquefying, which is Fischer-Tropsch synthesis. Following that, we delve deeper into the actual raw materials that can be liquefied by different means.

Fischer–Tropsch synthesis (FT)

With Fischer-Tropsch synthesis, it is possible to turn biomass (BtL), coal (CtL) or natural gas (GtL) into liquid fuels. During the Second World War, the Nazis produced 9 percent of their military and 25 percent of their transportation fuels in FT plants using coal and lignite as feedstock. Their peak production reached 124,000 bpd. After allied forces started bombing their FT plants, the Nazi war machine found itself slowing down. There are a few CtL and GtL plants operating around the world, but this chapter focuses more on biomass-to-liquids, with it being a next generation biofuel.

One of the main advantages of FT is that it is commercially proven to be somewhat viable. Also, it is not picky about the quality of its feedstock. It can be used for flexible production of high quality liquid fuels that can be used as drop-in fuels for our current fleet. Drop-in means that there is no modification needed. It needs to be remembered though, that FT is not an energy source. It is a method of transforming one kind of chemical energy into another kind of chemical energy. There are always losses, which are measured as the efficiency of the process. The theoretical maximum efficiency of a stand-alone FT plant using coal as feedstock is around 60 percent, with 50 percent being closer to real-world performance. With biomass, the efficiency drops to around 30–40 percent due to the poorer quality feedstock. VTT, the Technical Research Center of Finland, has said that with certain co-production methods it is possible to reach a total efficiency of 70–80 percent. When integrated with a paper or pulp mill, the waste heat and waste products can be efficiently used in the FT processes.

In short, the process has two phases. First the feedstock is gasified to syngas, which mainly contains carbon monoxide and hydrogen. The syngas is then washed to remove toxics, after which the relative amounts of carbon monoxide and hydrogen are converted to hydrocarbons. After this, FT synthesis is used to liquefy the gas, which is then refined into the desired products. The precise methods to be used with biomass are still under development, since there is not yet much commercial production.

Case: UPM biorefinery

UPM, the paper and pulp producer that turned into a "Biofore" company, has been planning a FT plant for some years. We use its proposed plant as an example here to find out what BtL can bring to the table. The proposed plant would produce 300,000 tons of refined products a year, mainly biodiesel and naphtha. It corresponds to around 6,000 bpd. It was planned to be integrated with a paper or pulp mill to take advantage of co-production opportunities. The plant, planned to be located in Finland, failed to get funding from the EU, while another similar plant planned for Strasbourg, France, secured 170 million euros from the European new entrants' reserve (NER) fund.

The planned refinery would use 1.5–2 million cubic meters of wood (mainly woodchip and such) annually (around 3–4 TWh). In addition, it would use 0.4 TWh of electricity and 5,000 tons of hydrogen. Other feedstock includes nitrogen, oxygen (400,000 tons per year), heat for drying, steam, cooling water and sweet water.[67]

What do these numbers mean? The EU uses around 13 mbpd of oil. As a "back of the envelope"calculation, replacing this oil with bio-oil is a daunting task. It would require around 2,200 such biorefineries, which cost 0.5 billion apiece. The refineries would require 880 TWh of electricity per year, which equals the combined electricity consumption of Germany and the UK, or around a quarter of the total EU electricity consumption.[68] The wood used as the main feedstock amounts to 4,000 million cubic meters annually. This is more than 6 times the annual growth of forests in Europe, and more than 10 times the current annual logging. The forests of Europe would be gone in a mere 6 years,[69] and the rather meager 1.3 percent slice that forestry-based activities contribute to European GDP would grow substantially, at least for a few years.[70] Of course this is only theoretical. On a more practical note, there have been fears that the price of wood that can also be used in paper production would increase with any large-scale liquefaction projects.

In addition to the refinery complex, it is also sensible to view the processes more holistically. If the electricity used in the processes was produced using biomass, the amount of biomass (wood) transported and used would rise by maybe 20 percent. It is worth also noting that around 5 percent of the energy content in the wood is used in its harvesting and transportation, but this energy is in a more valuable form as a liquid fuel, usually fossil. The hydrogen is normally manufactured from natural gas, another fossil fuel that is mainly imported to Europe, but it can also be taken from the process itself.

Is co-production essential?

Without the increases in total efficiency and other synergies that integrated plants bring, it would seem that the process makes little sense, at least from an energy-efficiency point of view. On the other hand, stand-alone refineries can be standardized more efficiently and are more flexible in their location.

The EIA has estimated (2006) the capital investment costs for different kinds of liquefaction plants on a dollars per daily barrel of oil produced:

- Gas to Liquids: 30,000 US$
- Coal to Liquids: 60,000 US$
- Biomass to Liquids: 120,000 to 140,000 US$ (this is more than 5 times the cost of a conventional oil refinery).[71]

In the summer of 2013, INEOS Bio finished the world's first commercial-scale cellulosic ethanol plant with a capacity of around 700 bpd.[72] Another of the world's first commercial-scale biorefineries, which could be built in Finland by VAPO Forest BtL, has an estimated cost of 500 million euros and a production capacity of around 2,000 bpd (100,000 tons per year). The investment per barrel of daily production is therefore around 250,000 dollars, around double what the IEA estimated.

The VTT Technical Research Center of Finland has developed processes based on pressurized fluidized-bed gasification that can be used to liquefy lignocellulosic biomass (such as logging residue and other forest industry waste). The five different process options are presented in their 2013 report "Liquid transportation fuels via large-scale fluidised-bed gasification of lignocellulosic biomass."[73]

Some of the processes have already been proven on a pre-commercial scale, and are therefore ready for commercial demonstration. Some of the processes, such as higher gasification pressure or higher temperatures, still need further development.

The report presented four different end products manufactured in a 300 MW (biomass usage) plant: methanol, DME (dimethyl ether), Fischer-Tropsch liquids, and synthetic gasoline (MtG, methanol to gasoline). The efficiency of biomass to liquids conversion varied between 50 and 66 percent. If waste heat was used for district heating, the total efficiency rose to 74–80 percent. The report concentrated on stand-alone facilities to make the energy and raw material flows more transparent. In practice, it would be sensible to integrate these plants with a paper mill or such. Not counting the higher costs for any prototype plant, the costs for these plants range from 320–420 million euros (roughly 430–550 million US$), depending on the desired end product.

Perhaps the most interesting thing about the report is the price of oil needed for these biofuels to be competitive without any tariffs or carbon taxes. For methanol, which can be mixed with gasoline in small amounts without the need for large modifications, the break-even price of oil is around 100 US$ per barrel. For synthetic gasoline the price is around 150 US$ per barrel.

The obvious problem for larger scale implementation comes from the availability of cheap feedstock such as logging residue. It is also possible that current societies cannot afford to pay much higher prices for oil (around 110 US$ per barrel for the last few years) without going into recession. On the other hand, biofuels will lessen the need for oil imports, which would leave the money circulating more locally, and presumably the carbon emissions could be much lower from these kind of biofuels compared with fossil oil.

Liquefied coal (CtL)

> King coal is extending his kingdom. So dysfunctional is the world's response to climate change that every year, the dirtiest fuel of them all is generating a growing proportion of the world's energy.
>
> Fred Pearce, 2011

By gasifying, then liquefying, coal with Fischer-Tropsch synthesis it can be transformed into high quality liquid fuels. By far the most common use for coal today is to burn it to create electricity and heat. Coal is often linked with steam engines, steamships, and the famous London smog of 1952, which killed approximately 12,000 people. It was the primary fuel for early industrialism, and that can almost be said for modern industrialism as well. The sad truth is that we have never used as much coal per year as we do today. China uses around half of the world's coal, and new coal power plants have been popping up there almost weekly. Even the poster nation for renewables, Germany, builds more new coal plants every year.

Coal is the largest single emitter of greenhouse gases we have. As a result of mining and burning it, hundreds of thousands of people worldwide die prematurely. It was estimated that in 2010 alone, 13,000 people died in the United States as a result of small particles emitted through burning coal.[74] If we compare the lethality of our energy sources, coal is the undisputed number one killer, with nuclear being the safest.[75]

Coal resources

> Whenever coal reserve figures are updated, the estimates are usually revised downwards: estimates of world reserves (79% of which are held in the United States, Russia, India, China, Australia and South Africa) were decreased by more than 50% in 2005, to 861 gigatonnes.
>
> James Murray and David King (2012), *Nature*, 481, 433–435

It is hard to get reliable information on proven coal reserves. This is mainly due to lack of reliable estimation methods, and the rarity of estimates done. So far, everyone has slept well on the information that we have a lot of the stuff. But how much is a lot exactly? At what price? Where? Of what quality? When will we hit a peak in coal production?

In his book *Blackout* (2009), Richard Heinberg presents some studies that have been made on the subject. The trend seems to be that we have been moving from high quality coal that has been easy to mine, to lower grades that are in harder places to reach and deeper underground. Peak coal has been estimated to hit somewhere between 2025 and 2050. The alarming thing is that almost every time new estimates on reserves have been done, the amount of reserves has been downgraded, often by a lot.

If we start to replace crude oil with liquefied coal, even vast reserves start to shrink fast. We use almost 90 million barrels of oil per day. If we were somehow to liquefy all of our annual coal production, it would count for around half of our oil use. This is not possible in reality, since we don't have the CtL plants, and coal is needed for electricity production. Since coal is less energy-dense than oil, and around half of that energy is lost in the liquefaction process, any significant liquefaction projects would also see our coal reserves shrinking much faster.

It is slow and expensive to add coal production and build the needed transportation infrastructure to new deposits and areas. Lack of infrastructure is one of the biggest hurdles in increasing coal production, since it has to be built anew each time a new, more remote deposit is taken into production.

Coal to Liquids processes

There are three processes to liquefy coal: pyrolysis, direct coal liquefaction (DCL), and indirect coal liquefaction (ICL). Fischer-Tropsch is an ICL-process. Of these, DCL and ICL are the most useful in producing high quality fuels for transportation. DCL efficiency is around 60–70 percent while ICL efficiency is around 50 percent.[76] ICL however offers a more flexible variety of refined products. Not to go too deeply into the chemical processes involved, the processes are as follows: With DCL, the coal is turned directly into a liquid by adding pressure and hydrogen until the carbon molecules start cracking into smaller, liquid hydrocarbons. With ICL, the coal is first gasified, washed for impurities, and then liquefied.

Historically there have been two main occasions of commercial scale CtL operations. The first was during the Second World War in Germany and the other was in South Africa during apartheid. Both were built to secure domestic supplies of liquid fuels, and the plants in South Africa are still operational. The operator of those plants, Sasol, is currently the undisputed leader in ICL production globally. It produces around 30 percent of South

Africa's liquid fuels (180,000 bpd). In China it produces around 20,000 bpd, and 10,000 bpd is produced in the rest of the world.

Sasol also offers the only commercial-scale data on coal conversion efficiency. It produces 1–1.4 barrels of liquids per ton of coal. It is substantially less than what the EIA has estimated possible (1–3 barrels per ton), even though the coal Sasol uses is of high quality.[77] If 10 percent of current coal production was to be liquefied, we would get just a few million barrels per day of oil. This would increase the price of coal, which would cut into the profits and tighten coal markets.

In 2006, China was planning twenty-seven CtL plants and thirty other coal refineries (to produce methanol and other chemicals). In 2008, all but two of the CtL projects were frozen.[78] In 2011, one of the remaining projects, an 8–10 billion dollar plant capable of 94,000 bpd production, was also frozen.[79]

In the U.S., The National Petroleum Council has proposed increasing CtL production to 5.5 mbpd by 2030. This represents more than a quarter of current U.S. oil use, but it would use more coal than is currently produced in the U.S. It does not sound very realistic.[80]

The price and EROEI of production

Coal remains relatively cheap as a feedstock. The problem with CtL is the substantial costs to build the facilities. One CtL plant costs at least hundreds of millions, and the project can last for several years with all the planning and permits. A large-scale CtL program would take decades to get going, if it was ever attempted.

Coal mining and transportation are substantial users of diesel oil and fuel oil. Transportation accounts for around a fifth of the price of coal, so the price of coal is somewhat sensitive to the price of oil as well. The price also varies significantly, even fourfold, depending on the quality of the coal.[81] Massey Energy Corporation, one of the biggest producers of coal in the U.S., produces around 40 million tons of coal a year and uses around 40 million gallons of diesel oil (around 151 million liters).

How much of the energy in the coal ends up moving the car after liquefaction? If the EROEI of coal starts at 50:1, the liquefaction process takes around half of this. Refining, transportation, and storage take another 10 percent. The internal combustion engine of the car loses another 75 percent as waste heat. We end up getting less than 10 percent of the original energy in the coal to actually move the car. We would get 2 to 3 times more movement if we simply burned the coal for electricity to power an electric car in the first place.

Environmental footprint of CtL

When first gasifying and then liquefying (ICL) coal, most of the toxins can be removed from the fuel. We cannot capture the carbon dioxide, however, without building some kind of carbon capture and storage (CCS) facility. CCS takes a lot of energy (making the efficiency even poorer), demands huge investments in infrastructure to transport the captured carbon, and of course a suitable (preferably nearby) location in which to store it. The CO_2 emissions from CtL are around twice those compared with conventional oil. Both DCL and ICL use large amounts of fresh water, with a rough estimate being a cubic meter of water for one barrel of oil (1:6).[82] This is substantially more than what is used in producing bitumen from oil sands (1:2–4.5).

Biofuels

> There is no yield. The literature said that with dry land, after four years' growth, you can get a yield of 1kg per plant. For us it is hardly 200g per plant.
>
> Professor R.R. Shah on the results of their jatropha planting without fertilizer[83]

Biomass is the only renewable energy source that can be turned into liquid fuels with reasonable costs and efforts. But there is biomass, and then there is biomass. The usual suspects in biofuel conversations are crops such as wheat and corn, oil plants, wood and other cellulosic plants, agricultural waste, food industry waste, and several others. Peat is also a biomass in this sense, although it is not renewable in the same sense. It does not grow, but it does accumulate in natural state peatlands. Algae have also been mentioned as a future source of liquid fuels, but they are still in their development stages.

The mental image of biofuels is a powerful one and for most it has very positive connotations. Most of us would like to cut our reliance on foreign oil and drive to work with liquefied domestic wood chips, or even waste, in our car. However, biofuels encompass a wide variety of feedstock and processing methods. Some of these are more sensible and efficient than others, so it is unhelpful to talk about biofuels in general as a good or a bad thing. For example, ethanol can be made from waste from the food industry, which is a sensible thing to do. The problem is that there is a limited supply of suitable waste. Another problem is that in order to produce that waste, a substantial amount of oil and other energy sources has been used, so its sustainability is questionable. Waste is not an energy source. We cannot move to waste-ethanol even if we would like to. But still, it is probably sensible to use that waste to make ethanol, especially if there are no other uses for it.

The main ways to produce biofuels today are:

- extraction from various oil plants (first generation)
- fermenting sugars (wheat, corn, agricultural waste) into ethanol (first generation)

- gasifying and then liquefying them from biomass with BtL processes such as Fischer-Tropsch or pyrolysis (second generation).

Other third and fourth generation processes still under research and development include:

- different ways of cracking cellulose into sugars, for example with the aid of bacteria
- biofuels from algae
- direct conversion of sunlight and CO_2 into liquid fuel.[84]

These processes and fuels have numerous things we need to consider when estimating their viability:

- Efficiency of the process. How much external energy and feedstock does it require, and how much energy we are left with?
- The quality of the feedstock. Is it edible for humans? Does it require high quality soil that could also be used for food crops to grow on? Are there other valuable uses for the feedstock and what are the costs of substitution?
- Is the process viable on a commercial scale, or is it something that can only be done in the laboratory?
- Can it be deployed in different kinds of locations on a commercial scale?
- What is the environmental footprint, and are there other hazards of production?

Currently, biofuels are manufactured to be drop-in substitutes for their fossil counterparts. In the future, there might be other alternatives to bioethanol and biodiesel that can be manufactured with different processes and from different feedstock. These include methanol, butanol, propanol, dimethyl ether (DME), and ammonia.

Growing biomass

It takes immense amounts of land to grow biomass for biofuels. It is more land-intensive than any other method of energy production. There is always the opportunity cost, since this land cannot be used for other purposes, such as growing food for humans, or preservation of ecosystem biodiversity. With a growing population on the one hand, and loss of fertile land due to erosion, urbanization and other infrastructures on the other, it gets harder and harder to feed the global population, let alone grow biofuels to power our ever-growing transportation fleet. Around a third of our ice-free surface area has already been directed towards human uses (12 percent for agriculture and 22 percent for pasturage).[85] From this point of view, first generation biofuels from food crops make little sense.

As an example, we can look at the amount of work and fossil energy inputs in the growing of corn-based ethanol in the U.S.:

Phase 1 – agricultural production of corn: 12 h of labor and 29 GJ of fossil energy/ha/year.

Phase 2 – fermentation/distillation of ethanol: 15 h of labor and 32 GJ of fossil energy/ha/year.

The energy output in ethanol is around 66 GJ/ha/year, which gives us an EROEI of 1.1:1. The net amount is around 277 liters of ethanol/ha/year.[86] The average power of ethanol production per hectare amounts to around 200 watts. The typical average constant power of transportation per person in industrialized countries is 1,000 watts. This means that, theoretically, if we were to grow corn ethanol to use as our transportation fuel, every person would require 5 hectares (12 acres) of prime land just for that purpose. If we divided our landmass equally among humans, we would get less than two hectares (5 acres) per person, and that includes deserts, glaciers, and such. There is simply no sense in this type of biofuels production.

Second generation biofuels that are cellulose-based are a step in a more sensible direction. They can be grown on more marginal land, and they can utilize many waste products as feedstock. One problem does remain, however, as land and its production is once more used by humans at the expense of biodiversity. But it does require less energy inputs in the form of fertilizers and soil cultivation. With sensible and more sustainable land use, a more vegetarian diet, and more local food, many of these problems could be mitigated. There have not been many signs of OECD citizens voluntarily cutting back on their use of meat, although some trends can be seen, for example in the U.S.[87]

The energy use of transportation is immense. In 2008, Virgin Airlines, owned by Richard Branson, made a test flight from London to Amsterdam using 20 percent biofuel made from coconuts. The flight used 150,000 coconuts.[88] With a quick calculation, it would take around 12 million coconuts to make an all-biofuel transatlantic flight from London to New York. By using our total harvest of corn and sugar cane as ethanol, and also including biodiesel produced from our total harvest in soy beans and palm oil, we could have replaced around 10 percent of our total liquid fuels used in 2006. Turning agricultural waste into biogas would have replaced around 26 percent of the energy used in the transportation sector.[89]

Global production capacity

Regardless of the relatively large media hype, according to the IEA, biofuels won't have a very large impact on humanity's liquid fuel use during this decade. The global capacity will stay at around 2.5 mbpd by 2020.[90] As ethanol only has around two-thirds of the energy content of crude oil, the actual amount is equal to 1.7 mbpd of oil.

The production capacity is expected to grow around 100,000 bpd every year, which is around 65,000 bpd in oil equivalent. It can offset around 1.5–2 percent of the decline in current fields. The majority of current production and future growth comes from corn ethanol in the U.S. and sugar-cane ethanol in Brazil. The drought in 2012 opened some discussions in the U.S. about the sense in growing corn just to make liquid fuels.

Problems with biofuels

A lot of problems are associated with biofuels, and a lot of them for good reason. Some of the problems only manifest themselves with certain types of biofuels, but many have wider negative consequences. The subject is a controversial one with environmentalists and Greens, since it was the Green Party in Germany that was instrumental in the birth of the EU's current biofuel mandates. In the autumn of 2012, a draft was leaked that suggested some gaps in the ambitious aims. It was seen that the 10 percent biofuel mandate was contributing to global hunger, increasing food prices and had little effect on emissions due to land use changes. The draft suggested that the amount of biofuels should perhaps remain at 5 percent by 2020.[91]

The costs for producing biofuels rise rapidly as we move away from first generation fuels such as ethanol and palm oil. The costs of producing a second generation cellulosic biofuel can be many times higher than their first generation counterparts. To counter these expenses, these biofuels often require a host of public monetary and regulatory support. This support has been forthcoming based on the perceived emissions cuts. As the reality of the situation, where there is little, if any, actual cuts in emissions, spreads among politicians, and citizens, it will become harder to justify the support.

Most current methods of producing biofuels use land that could be used to grow food. This creates an incentive to chop down forests and drain wetlands for agricultural use, either directly to grow biofuels or indirectly to grow more food on it. As the changes in land use can increase the emissions coming from the soil, the total emission cuts from biofuels can in fact turn into increases in emissions. Agriculture is already one of the biggest emission sectors we have. Biofuels often mean that we are cutting emissions from one sector (e.g. transportation) and adding them elsewhere (e.g. agriculture).

Annually planted field crops especially demand a lot of fertilizers, herbicides, and other products that are mainly manufactured from fossil fuels. They also demand yearly harvesting and transporting, which is also done with machines running on fossil diesel oil. Often these inputs are omitted from the calculations, as producers and companies demonstrate their products, efficiencies, and emission reductions. Due to the massive use of fossil inputs, biofuels produced on more marginal conditions can even have negative net energy output. When this happens societies are left with less energy, but as a consolidation they can marvel at the rising volumes of liquid fuel production. As a whole, this does not make much sense.

Perennial, usually cellulose-based, plants such as trees demand much less energy input, as they are usually harvested less often. Harvesting energy wood, at the same time as normal chopping, uses 3–6 percent of the energy content in the wood collected.[92] It should be noted that this energy is used as highly refined liquid fuels, while the wood collected is much less valuable as an energy source.

Social implications

Biofuels are often defended in western countries by saying that they will offer people jobs. The situation might be totally different in developing countries, where the business is firmly in the hands of big multinational corporations. Sugar cane in Brazil is cheap and energy-efficient, as the harvesting is largely done manually (with less-than-stellar salaries for the harvesters). In the Far East the land of small farmers is grabbed by their government and corporations, forests are cut down, and palm oil is planted instead. The original lifestyle and subsistence of the locals disappears into the tank of a western Mercedes, seemingly without a trace.

It is a well-known fact that as oil prices increase, so does the price of food, as inputs and transportation become more expensive. Biofuels bring another component to the pricing table. As the price of oil increases, it becomes more profitable to make substitutes such as biofuels. This increases the pressure to produce biofuels. The World Bank published a report in 2008 stating that biofuels were responsible for 75 percent of the price rises in food.[93] More expensive food punishes the poor, who already use most of their income for food. Many experts have predicted worsening food riots in the coming years. They say that we have around a year until food prices go over the critical limit.[94,95]

Transferring most of nature's primary production to humans

Increasing biofuels and bioenergy production in general will lead to an increasing part of nature's primary production to be directed for humanity. This will have dire ecological implications. As more and more is directed for use by humanity, less energy is available for the rest of the ecosystem. This will eventually lead to ecosystem collapse. As the earth's ecosystems are the ones producing the energy and materials in the first place, with the disappearing ecosystem it will crash humanity's ability to get its former share. A diverse ecology and thriving nature offer us many valuable and even invaluable services and products such as fresh and clean water, oxygen, and pollination of plants and crops.

Many agricultural practices lead to soil erosion in time. Producing biofuels may be self-defeating in the longer run, as the marketplace encourages short-term maximization of production, and regulation is often lacking, distorted or both. Modern agriculture is responsible for 75 percent of biodiversity losses. We are experiencing the sixth extinction wave, which many think is a comparable threat to climate change. With biodiversity we also lose the flexibility and resilience of our ecological systems, which could turn out to be an expensive and devastating loss with ongoing climate change.

Summary

Biofuels will play their part in our future transportation. Currently they often end up being greenwash for companies. As the price of oil climbs, they can also become a business opportunity. The multinational corporation UPM seems to have come to this conclusion. We tried to interview the company, but they cancelled the interview after receiving our questions, claiming that their operations were so new and small that they did not have much to say. A few weeks later, UPM's CEO Jussi Pesonen was interviewed in a Finnish economics publication *Talouselämä* with the headline: "This will be a business in the billions of dollars." In the piece he said the following:

> There will be scarcity of resources in the global economy. We need to use more renewables and recycle more efficiently, with more value produced . . . The golden age for all biofuels will come when crude oil hits 200 to 300 dollars per barrel. This can take one or two more recessions, but the time will eventually come.

It is no big surprise that the CEO of a large company is more aware of the coming megatrends than most politicians. A businessman primarily sees the business opportunity in the situation, and only secondarily any threat to the stability of the society. It is however possible, even likely, that the depletion of oil and other fossil energy sources will cripple the whole of society in ways that are hard to see from the perspective of a CEO. As increasingly expensive oil drives the global economy towards more and more instability and makes growth harder, it will also have negative effects for the balance sheets of most companies. Also, if we face a more sudden oil crisis, there is an incentive for governments to declare a state of emergency, which will give them the power to nationalize companies and their assets, for example.

If our aim is to be a carbon-free society, or even to achieve meaningful reductions in carbon emissions, then most of the current biofuels are unsustainable. Their production and use often causes indirect emissions that are comparable to any reductions at the tailpipe. When peak oil is identified as the challenge to overcome, along with goals of oil independence and greater self-sufficiency for a nation, biofuels make more sense. The global situation will in the future become more like a wrestling match between mitigating climate change and ensuring energy security.

Biofuels have so many differences that, as mentioned, declaring them categorically as a good or a bad idea needs to be avoided. Constructing plants to produce waste ethanol, biogas and maybe some synthetic bio-oil from Fischer-Tropsch synthesis might be reasonable ideas. Biofuels based on algae and other later generation processes might end up being viable options, when they eventually get to market.

One thing is painfully clear though. Biofuels will not be able to supply even close to our current oil use; not for decades, if ever. Cellulosic biofuels need expensive manufacturing plants that we do not have, and their prices will follow the prices of oil. Even our current global biofuels production causes rising food prices and civil unrest. We will have neither sustainable nor rapid growth of biofuels in the coming years.

Transportation without oil

The best way to help with society's oil dependency is to cure the disease. The easiest way is to give up activities that use a lot of oil, such as our current driving habits. If this is not possible, or is seen to be otherwise undesirable, we need to replace petroleum-based fuels with some other energy. This means that our vehicle fleet needs to be replaced, or modified at least partly, and the distribution and storage network for the replacement fuel needs to be built. Below we look at three different fuels that can be used to do this: using natural gas as transportation fuel, electrifying transportation, and using hydrogen as a fuel.

Natural gas can be utilized with the least changes to the current fleet. The internal combustion engine remains the same, so it is possible to modify current vehicles to use natural gas. Further, it is possible to liquefy natural gas in Fischer-Tropsch plants, when it can be used as a drop-in fuel for the current fleet. Electrifying our transportation calls for replacement of our current fleet, even though some degree of conversion is possible also to current vehicles. The distribution network is already in place, although it needs to be strengthened in many places if electric cars proliferate, as public charging stations are scarce. Further, electricity can be produced in a wide variety of ways, also without emissions. A hydrogen economy or even hydrogen-based transportation is farthest from realization. It requires totally new vehicles, and a totally new production and distribution network with a multitude of technical problems and bottlenecks.

Natural gas and other gases

There are a number of ways to produce natural gas or similar gases. From the viewpoint of their end use as a transportation fuel, the production method does not matter much. Gas condensates, liquid petroleum gas (LPG), and natural gas liquid (NGL) often mean similar things. NGL often means "wet gases" that are normally a by-product of oil drilling. They are relatively easy to compress to liquid, even if they are gases under atmospheric pressure. They include ethane, propane and butane. Gas condensates are usually a bit heavier than butane. Then there is LNG, where liquefaction is done by cooling the methane.

Conventional natural gas, of which the majority is methane, is the most significant of these gases. It is our third largest primary energy source (after oil and coal) with a share of over 20 percent. This share is estimated to rise to over 25 percent by 2035.[96]

Many environmental organizations have historically viewed natural gas as a reasonably benign and clean fuel. This is probably due to the clean burning of methane – there is little in the way of air pollution or small particles released. In addition, it releases only around half of the carbon dioxide compared with coal. When burned to generate electricity, the efficiency is relatively good at 40–50 percent. Gas turbines are also quick to fire up and easy to use for load following, which means their output can quickly be adjusted to meet electricity demand. This is a useful property if we aim to increase intermittent electricity production from solar photovoltaic panels and wind turbines. The problem is fugitive gases that are leaked during production, transportation, and storage. Methane is 60 times more potent as a greenhouse gas than carbon dioxide on a 20-year timescale (which is the timescale we need to accomplish serious reductions in emissions). If there is only a few percent of leakage overall, the comparable emissions of natural gas climb to the same levels as coal, or worse. This is painful for many to admit, as natural gas has played a major part in emissions reduction scenarios, especially those that have avoided increased use of nuclear power. According to some sources, there is a new slogan among industry and policy makers in the European Union: "Gas is the new oil." Emissions reductions in Europe have largely been shelved to wait for the economy and competitiveness to improve. The shift to natural gas is hindered by the lack of sufficient storage facilities. The difficulty of gas storage is a strategic weakness, as oil is much easier and cheaper to store for the rainy day.

Conventional natural gas

Often natural gas is the by-product of oil. Since it takes a lot of space and is hard to store and transport it was often released into the atmosphere a few decades ago. As the repercussions started to be understood, we started to burn it in gas flares. Even today, an amount equaling a third of Europe's gas consumption is flared globally. If the gas field is large and transportation can be arranged, gas is collected since it is a valuable fuel and feedstock for industry. The gas markets however remain more local than the global oil market, as gas remains harder to transport for long distances.

The production profile of gas fields is different from oilfields. Oilfields often see a few percent decrease after their peak. Gas fields, however, can see much more significant reductions in post-peak production. This makes peak gas harder to estimate. In addition, natural gas is often flared or used locally for energy without measuring, so the actual production amounts remain a bit vague. This information would make it easier to calculate future production from the remaining reserves. Some estimates[97] put peak conventional natural gas inside 10 years. The EIA predicts growth at least until 2030.[98]

The emissions from conventional gas depend on the production location and its geology, practices used in production and the shape of the transportation infrastructure. Even though its carbon dioxide emissions are half of those from coal burning, even small leaks of methane can negate that. So with natural gas we can reduce our emissions on paper, but in reality those emissions can remain. When considering emission targets or carbon markets, only the reductions on paper matter. When considering actual climate change, only the real emissions matter.

Natural gas in Europe

The European Union is dependent on imports for around two-thirds of its natural gas consumption, and around 84 percent of its oil consumption. As Germany is planning to shut down its nuclear plants, and with the banning of shale gas fracking in some European countries, these imports are set to rise rapidly. Most of the gas imports come from Russia (34 percent) and Norway (35 percent). Other major sources are Algeria (14 percent) and Qatar (10 percent).[99] European (EU-27) production seems to be in steep decline, with 2011 to 2012 production dropping by 5.5 percent.[100] European reliance on Russian natural gas is seen as problematic by many. Russia has shown that it is willing to use its energy might as a political tool, and has been active to prevent the situation from changing. As the Russian gas flows mainly through Ukraine and Belarus, the hostilities between those countries and Moscow have resulted in supply interruptions (2006 and 2009). Russia's natural gas production has been on a plateau since 2006.

According to a recent report "Europe's Energy Security: Options and Challenges to Natural Gas Supply Diversification,"[101] there are a few possibilities for the EU to diversify its future natural gas supplies as its hunger grows but production declines. The building of the Nord Stream gas pipeline between Russia and Germany is one way to contain political risks, but it does nothing to lessen Russian influence in Europe. The pipeline has a capacity of 2 tcf/year and is operating currently at 80 percent capacity, supplying more than a third of Germany's gas imports. Russia is also pushing for a South Stream pipeline across the Black Sea, which is seen by many as competition to the EU's other pipeline projects. This would help Russia to keep its grip firm on the European gas market and would also give it political leverage both in European countries and countries that the new pipelines bypass (Poland, Lithuania, etc.).

A second response is what is called the South Corridor, which aims to bring gas from the Caspian region and Central Asia to European markets. In June of 2013, the consortium controlling Azerbaijani natural gas chose a project called TAP (Trans Adriatic Pipeline that moves gas to Italy) to connect to TANAP (Trans-Anatolian natural gas pipeline). Both of these pipelines are significantly smaller than the formerly planned Nabucco project, which was not seen as commercially viable in spite of both U.S. and EU support. Their anticipated date for entering service is in 2019.

A third aspect is to integrate and liberalize the now fragmented European internal energy market, which the heads of states have pledged to do by 2014 (although officials say that this date is unlikely to be met). Further, the aim is to coordinate external energy policies, boost energy efficiency, and connect all member states to a Europe-wide energy supply network. The European Commission has put a price of around 1 trillion euros (1.3 trillion US$) on the needed investments.

Other competitors or options for Russian gas include North African gas supplies, though the region currently suffers from political unrest and poor infrastructure in the aftermath of the Arab Spring, among other things. And then there is the option of increased LNG (liquefied natural gas) imports from the world market. Currently LNG accounts for around 20 percent of European imports. Usually LNG facilities lie underused most of the time, as capacity is reached only in peak demand during winter. Instead of new LNG terminals, or in addition to them, facilities to store natural gas could be built to put current capacity to better use. There is also the possibility of growing LNG exports from the U.S., but that is often seen as more of a bargaining tool against Russian long-term contracts that price natural gas against oil prices. It remains to be seen whether the United States is both able and willing to export a significant part of its shale gas in the future, as it could be more valuable as a raw material for the U.S. chemical industry.

Russia, regardless of any developments to a more unified energy policy in the EU, will most likely continue to control European energy security. Its biggest customers, Germany and Italy among many others, have signed long-term bilateral deals with Russia that partly undermine any unified response. Gazprom, the government-controlled Russian energy giant, is shopping for ownership of storage and distribution infrastructure around Europe.

Russian point of view

If Russian gas imports are essential for Europe, so are gas exports to Europe essential for Russia. The EU is the largest customer for Russia's gas, and one that also presumably pays its bills on time. There have been problems with the Commonwealth of the Independent States (CIS) not paying on time and also getting subsidized prices. For example, Russia and Ukraine agreed to extend the stay for the Russian Black Sea fleet in the Crimea from 2017 to 2042 in return for a 66 percent discount on natural gas for Ukraine. In early 2014, the political situation in Ukraine became unstable, and Gazprom revoked the discount.[102] Given the wide gap between natural gas spot prices and prices pegged to oil prices, the discount has ended up as somewhat disappointing for Ukraine. The disputes between Ukraine and Moscow (mostly over money) have been a headache to the EU. Before Nord Stream, about 80 percent of Russian exports to the EU went through Ukraine. In 2006 and 2009 these led to Russia cutting its gas transport through Ukraine, leaving the EU stranded in the process. Gazprom has been aggressively trying to get control of the Ukrainian pipelines, while the Ukrainians would like to get discounted gas and guaranteed throughput volumes in exchange for jointly operated pipelines. The negotiations have been colored by Gazprom's recent bill of 7 billion to Ukraine for the natural gas that they failed to buy (in accordance with their "take-or-pay" contract).

Russia is the holder of world's largest natural gas reserves (18 percent of global reserves), the largest exporter, and the second largest producer (after the U.S.). It is also utterly dependent on income from its natural resources exports and taxes. At least half of the government's income comes from oil and natural gas exports. Gazprom is currently responsible for all of Russia's gas exports, of which half goes to the EU. The rest goes primarily to CIS countries and Turkey (which is applying for EU membership).

As the EU is looking for ways to diversify its natural gas imports, so Russia is planning to diversify its customer base. Currently only 7 percent of exports go to Asia, of which all is made in LNG. By 2030, Russia is planning to increase this portion to around 20 percent. The process of increasing exports to China by building pipelines has so far been halted by China's unwillingness to pay comparable prices to Russia's EU customers. Given this, and the immense costs of exploring and increasing gas production, the current situation will likely stay the same for years, if not for decades.

Gazprom has traditionally pegged its gas prices to oil prices. Recently, as the natural gas markets have been shifting, Gazprom has changed part of its pricing to spot-pricing, which offers much lower prices. This has been partly seen as a response to U.S. plans to export shale gas to Europe, where the prices have been much higher because they have been pegged to oil prices. Even if none of the U.S. shale gas ever reaches European markets, its effects have already started to show.

Shale gas

> The steep production decline rates of shale wells, which can reach up to 80 percent in the first year, and the advanced age of many of America's largest gas producing basins, have created a production treadmill that is likely to result in a natural gas deliverability crisis in the 2013–2015 timeframe.
>
> Bill Powers, *Cold, Hungry and in the Dark* (2013)

Much of what was said about shale oil earlier also applies to shale gas. Shale gas has overtaken around 40 percent of U.S. gas consumption in 10 years, and has made the U.S. all but independent on natural gas. This is a significant accomplishment, made possible by increasing natural gas prices, loosening regulation and advanced production methods. Hydraulic fragmentation and horizontal drilling, combined in fracking, have made it possible to produce natural gas and oil trapped in tight shale formations. How profitable this production has been and will be in the near future is a debate with widely varying views. Many sweet spots have been profitable, but it is a fallacy to assume this can be repeated after the sweet spots have been drilled.

The main problem with shale gas is that its potential production at any given price is still widely unknown and untested. Despite this, many enthusiastic decision makers are eagerly putting our long-term energy policy on assumed large increases in shale gas production. The assumptions have largely been made by the same industry that has been in need of investment capital and a greater fool to buy their leases and land, so it would be wise to take their message with a pinch of salt.

The actual money has been made on Wall Street. It worryingly resembles a pyramid scheme: land leases, drilling rights, pumping corporate value and flipping all of them over to the next fool willing to pay an even higher price.[103] Often the next fool is only buying, not because of any actual gas, but because they think an even greater fool can be found down the road. During the last few years, many corporations have been forced to write-off losses from reserves and assets worth billions of dollars.

A big part of such write-offs has been the crash of natural gas prices that have resulted from a gas glut on the market. There has been a frantic search for new uses for the gas to get the price back up. But why would the industry drill itself into a hole like this? Arthur Berman, an oil and gas analyst, suggests the following factors:[104]

- drill-to-hold type of leasing contracts that require the company to drill inside a certain time period if it wants to hold the lease on the land (no matter what the price of gas);
- the aim to sell the leasing contracts or even the company itself to foreign investors/oil companies that want to get into the shale game;
- the need to preserve the stock prices of the company by growing reserves and production;
- contracts of production minimums and building infrastructure such as gas pipes and joint-ventures with other companies;
- the wake-up to the reality that large global projects are getting scarce;
- predicting and trusting that natural gas prices will rise in the future.

Total natural gas production in the U.S. has been stagnant since late 2011, largely due to the rapid decline of drilling rigs. The gas glut that has enabled the low prices will be a short-term phenomenon. As the shale gas wells experience very rapid depletion, there might even be a gas deliverability crisis in the works in few years' time. In August 2013 there were only 386 rigs drilling for dry gas, which is a 78 percent drop from the all-time high of 1,606 in 2008. From the 2012 average, the drop is 109 rigs.[105] Shale gas is still shaking the world. In 2012 Gazprom pulled the plug on its $15 billion flagship project, Shtokman field in the Arctic, citing rising costs, falling European demand and cheap U.S. shale gas as the main reasons.[106] Falling European demand, at least in the longer run, is in conflict with the report presented earlier in the book, and cheap U.S. shale gas is a short-term phenomenon.

The estimated reserves in the U.S. have also seen drastic reductions at the hands of more unbiased geologists. The Marcellus formation had an 80 percent drop in reserves in the fall of 2011. The former estimates had been done by consultants close to the gas and oil industry and the methods were never published. It remains to be seen what will happen to the total reserves. These reserves depend on the price of natural gas. The break-even price of current shale gas operations have been estimated to be between 5 and 8 US$ per thousand cubic feet (mcf).

The much-hyped 100 years' worth of gas (2,170 bcf) in the U.S. can be roughly categorized as follows:[107]

- 11.4 years of proven reserves;
- 22.4 years of probable (but unproven) reserves;
- 28.7 years of possible reserves from undiscovered fields;
- 21.6 years of speculative reserves. Somebody thinks that somewhere there might be this amount of gas;
- 7.3 years of coal bed methane (of which around 10 percent are actual proven reserves);
- the remainder from a rounding error, to get the nice, round number of 100 years.

It would seem that one could also have rounded the numbers quite differently. On the other hand, there are even larger estimates, such as those put forward by billionaire T. Boone Pickens.[108]

Shale gas around the world

> We know the gas in place is there, but it's unclear if it can be extracted at a rate that's commercial.
>
> Laura Loppacher, oil and gas analyst on Polish Shale gas resources in 2012[109]

There are shale gas resources outside the U.S. In Europe, France and Bulgaria have banned shale gas drilling. Poland recently saw its resource estimates come crashing down, as the highest EIA estimates were lowered by more than 90 percent. Poland might hold gas for decades of domestic consumption, but it will be expensive to produce with little chance for significant export revenues. By the summer of 2013, half a billion in exploration investments had been made and forty-five test wells had been drilled in Poland. None of them have been commercially viable. This is depressing news for Europe, since Poland has been seen as the most promising source for shale gas.[110] In 2012, China published its own estimates on its shale gas resources, wiping the table with double the shale gas compared with the United States.

In its 2013 report[111] the EIA gives some preliminary estimates on unproven shale gas resources. Shale gas is estimated to increase total resources by 48 percent, making shale gas count for around a third of total global resources of 20,451 tcf. According to the report the top ten shale gas countries are:

1	China	1,115 tcf
2	Argentina	802 tcf
3	Algeria	707 tcf
4	U.S.	665 tcf
5	Canada	573 tcf
6	Mexico	545 tcf
7	Australia	437 tcf
8	South Africa	390 tcf
9	Russia	285 tcf
10	Brazil	245 tcf.

It needs to be remembered that unproven resources are not reserves. To compare, we have vast resources of methane waiting for us in the atmosphere of Titan, which is a moon of Saturn, but these are not reserves as they cannot be produced profitably. The EIA reminds in its report that it is unclear at what price these resources can be produced, and also that they are preliminary and unproven estimates. There could hardly be more conditionals and uncertainties, but the media has largely ignored these. The result is a misinformed public and policy makers that base their decisions and plans on nothing but uncertainties.

The role of shale gas will be significant. But will it be game changing? How long will it last and at what price? Should we plan our energy future on something as uncertain as shale gas? Will it be a bridge fuel to low-emission nuclear and renewables, or will it be a bridge fuel back to coal? Or, as even coal starts to deplete, a bridge to nowhere?

Biogas

Biogas can be crudely separated into three categories: waste-based, agriculture-based, and wood-based. The first two categories include using bacteria to digest biomass to make biogas in bioreactors or collecting methane from landfills and wastewater treatment plants. Since the methane would leak from the landfill anyway, this kind of activity to collect it and burn it for fuel is highly recommended, but limited in scalability. Wood-based biogas is usually made by gasifying the wood chips with pyrolysis and similar processes.

Bioreactors can be of any size, and there are millions of farm-scale bioreactors around the world.[112] Biogas can be produced locally on a small scale, where the resulting gas can be burned for electricity and heat, or even directed to fuel small-scale transportation needs or to power farm equipment. This kind of activity can increase in popularity in a world of expensive oil. The best use for biogas is in transportation, since there it replaces fossil oil directly.[113]

According to a 2008 study[114] the theoretical biogas potential from the U.S. meat industry is around 300 TWh per year. It is around 1 percent of the nation's total primary energy use. According to the EPA (Environmental Protection Agency),[115] 8,200 milk and meat farms in the U.S. could collect around 13 TWh of biogas, which would equate to the capacity of two 800 MW coal-fired power plants. The difference between the theoretical potential, calculated using the methane production per 1,000 pounds of animal, and the more realistic actual potential given by the EPA is huge.

Where the potential of biogas from the meat industry is rather limited, the benefits are double from the point of view of greenhouse gases. The methane would otherwise leak to the atmosphere, and by collecting it we can burn it for energy instead of another (mainly fossil) fuel. The energy produced by the animals in the form of methane does not come from nowhere, however. Eating less meat would be far more useful from the point of view of both energy conservation and emission reduction. The larger potentials for biogas lie in plant matter digestion in bioreactors and gasifying wood-based materials.

Condensates and natural gas liquids (NGLs)

A surprisingly large portion of our oil production is actually natural gas liquids. These are gases that are heavier than methane, such as ethane, propane, and butane. According to the IEA[116] more than half of our future growth in oil production will actually be from NGLs, and 90 percent of this growth will be from OPEC countries. This estimate assumes that the production of natural gas and shale gas will keep growing rapidly for the coming decades, as NGLs are often a side product from gas production. In short, if natural gas production does not grow, NGL production will not grow either, and total oil production will take a hit.

By volume, almost 10 percent of global oil production is natural gas liquids and similar products. Measured by energy content, they only hold 60–70 percent of the energy compared to the same volume of crude oil. It is therefore somewhat fallacious and dangerous to count them 1 to 1 with crude oil, which is the normal practice. This distorts energy policy.

Liquefied natural gas (LNG)

LNG is natural gas that is liquefied by sufficient cooling. Methane takes far less space as a liquid, so it becomes more economical to transport it by tanker ships, for example. At the turn of the century, many LNG import terminals were built in the U.S. as the nation's gas production was in decline. With the shale gas boom, these terminals are now largely useless, and there are talks to modify them for export. The economic and political web is complicated.

The gas producers would like to export gas to Europe or Asia, where the prices are many times higher. Depending on capacity, this would increase the price of domestic gas in the U.S., and lower the gas prices in Europe and/or Asia. The chemical industry is interested in cheap domestic gas, as it gives it a competitive advantage in the world market of chemical products made of gas and NGL. Politicians not in the pockets of the oil and gas industry would like the maximum value of the gas to stay in the U.S. This would mean refining the gas domestically, with all the jobs, foreign and domestic investments and added value staying in the U.S. There are problems here also, as the chemical industry is not located near to the new shale gas production sites. This adds costs if the gas is to be transported, since the infrastructure is lacking. On the other hand, local politicians would like the chemical industry to move to their state so the gas would not need to be transported anywhere, with the jobs and added value. It is not a big surprise that export licenses and the host of other permits are not easy to get.

The most interesting question is whether there will be any cheap shale gas to export in a few years' time when the export terminals are ready, if they are built? It would be a double irony if the U.S. would first build largely useless import terminals only to follow up with a fleet of useless export terminals a few years later.

The global LNG market might be growing, but probably not as much as many are predicting. The LNG not going to the U.S. anymore has found willing buyers in Asia, especially as Japan shut down its nuclear fleet after the Fukushima accident. China has also been growing its import capacity, which is estimated to grow tenfold between 2008 and 2015 from 0.5 bcf to 5.9 bcf per day. As discussed earlier, Europe is viewing LNG imports as one way to diversify its dependency on Russian gas, with the trading moving from long-term agreements and oil-pegs to spot trading. On the other hand, Qatar, which is the world's largest LNG exporter with a 31 percent share (10.3 bcf per day), is showing worrying signs. It also has the world's largest natural gas field, the North Field. In 2005 it banned all new gas projects. Since then it has directed its gas production more to the GtL-plant it built with Shell and its growing chemicals industry. The massive reserve estimates in the North Field have come under criticism for being way too optimistic.[117]

Finally, liquefying and transporting natural gas is not free. The cost to liquefy, transport gas to Europe and re-gasify it is around 4.5–5 dollars per mcf. The cost to Asian markets is 6.5 dollars. The estimated break-even price for shale gas is 5–8 dollars. This means that gas prices in Europe need to be at least 10–13 dollars if LNG is to be profitable. In addition to Europe and Asia, there is also a third competitor for U.S. shale gas: Mexico. Its gas imports from the U.S. have been growing rapidly: 92 percent from 2008[118] and 24 percent in 2012 alone.[119]

Compressed natural gas (CNG)

CNG is another way to store and transport natural gas. Its energy density is lower than that of LNG, but it doesn't have to be cooled like LNG. Most natural gas vehicles use CNG, which takes up considerable space in cars.

Gas to liquids (GtL)

Methane and other gases can be used as feedstock to make synthetic liquid fuels with processes such as Fischer-Tropsch. These fuels can be used as drop-in fuels for the current fleet, but a significant part of the energy content is lost. Sasol, known also for its coal-to-liquids projects is the global authority on GtL also. The current global capacity is around 210,000 bpd, with 180,000 bpd coming from the joint venture of Sasol and Qatar Petroleum and the rest coming from Nigeria.[120]

Gases in transportation

Natural gas (NG) has a few strengths and a few weaknesses as a transportation fuel. Methane is harder to transport and store than gasoline or diesel. The infrastructure for methane use is still lacking in many places. Storage facilities are more expensive to build and maintain than oil storages. Refueling locations are often few and far between. This has partly been circumvented by adding a small gasoline tank to natural gas vehicles.

There are thousands of public CNG refueling stations in Europe. Germany and Italy hold the most, with around a thousand stations in each. Other relatively dense networks are in the Netherlands, Switzerland, Austria, Sweden, and Bulgaria. Each of them has more than a hundred filling stations.[121] With expensive oil, locally produced and distributed natural gas can be a valuable asset to a community. Farm equipment such as tractors running on methane have also started to show up.

The emissions from NG vehicles are usually seen as lower than comparable gasoline or diesel vehicles. It is unclear however what the total lifetime

Energy content for transportation fuels (approximation, per liter)

Gasoline	9.5 kWh
Diesel	10 kWh
Biodiesel (vegetable oil)	9.2 kWh
E85	7 kWh
Methane (gas)	0.01 kWh
Methane/CNG (250 Bar)	2.5 kWh
Methane/LNG (–160°C)	6.1 kWh
Hydrogen (gas)	0.0028 kWh
Hydrogen (690 bar)	1.25 kWh
Hydrogen (liquefied)	2.35 kWh
Lithium battery	1–2 kWh

emissions will be, when leaks and such are also taken into account. Depending on the calculation method, biogas can have nearly zero emissions.

A gasoline vehicle can be modified to run on methane. The investment per vehicle is a few thousand euros. CNG at 250 bar pressure has around 25 percent of the energy content per volume compared with gasoline, so the trunk has to be partly sacrificed in the conversion. The conversion of diesel vehicles is harder to do. In addition to cars, natural gas, often LNG, can also be used to fuel large ships. One example is the M/S Viking Grace that entered service in early 2013.

Electricity

Electricity is not a replacement for oil, because it is not an energy source. Electricity needs to be produced with some other energy source, and electricity acts as a carrier of energy. For powering transportation, electricity does have some benefits compared with hydrogen. For example, it has a production and distribution network already built (if not always sufficient) in many places. On the other hand, the infrastructure of fast-charging stations is still weak. The EU is solving this problem with the usual means: a directive for cleaner transportation. According to this, Europe has to have 8 million charging stations by 2020, with 10 percent of those public.

The problem with electricity is that our fleet needs to be replaced, which is a project that will take decades. Another option to increase electric transportation is to build electrical mass transit (electric rail, tram-lines, and so forth). Indirectly it is also possible to replace liquid fuels with electricity within our current fleet. If we forget about efficiency and prices, it is possible to synthesize liquid fuels from water (hydrogen by electrolysis) and carbon by adding enough energy (electricity and heat). This hydrogen can also be used to make fertilizers and replace natural gas as their source of hydrogen. Anything like this requires that we have a significant surplus of cheap low-emission electricity production from renewables and nuclear. This is not the situation in most places, and not for the foreseeable future.

When these factors are taken into consideration, it is difficult to see how Europe could build an excess of low-emissions electricity production capacity, even if some energy-intensive industries were to relocate elsewhere. Low emissions electricity will find its users in a world of expensive oil and worsening climate change, although the transfer network might become a worsening bottleneck, as it already has in some places such as Germany. Synthesizing liquid fuels with FT-synthesis from biomass, for example, is an energy-intensive process.

Electricity production

Electricity is usually produced with different water boilers in which steam is used to spin a generator, which produces electricity. The fuels include coal, uranium and biomass such as woodchips. In combined heat and power (CHP) plants the waste heat is collected and used for different purposes such as industrial processes and district heating. This increases the total efficiency radically. This is a great thing, but it needs also to be remembered that building a comprehensive district heating infrastructure also locks the area into burning something for power and heat for decades forward. An area with district heating inoculates itself from the benefits of geothermal power (heat pumps), solar heating, and others, since individuals choosing these will lead the collective to use its infrastructure and power plants less efficiently. This means that the same investment in these alternatives would be much more beneficial somewhere else, where it would displace oil heating systems, for example.

Another significant method of making electricity is to use kinetic energy from wind or water to spin a generator. Hydropower is relatively affordable and reliable, especially after it has been built, but the local effects are less pleasing. The dam area covers large areas with water, displacing man and beast alike. The rotting vegetation also produces methane emissions, which is a significant problem in tropical zones. Some cases have resulted in worse emissions than if the same electricity had been produced with oil. Hydropower is most likely not as low emission electricity, as has been commonly thought.[122] Dams also prevent salmon from traveling upriver to spawn, which affects the local fishing industry and tourism.

In the developed world, the best places have largely already been built, so it will be hard to add hydroelectricity capacity. The global estimated technical capacity is 16,400 GWh per year. The largest untapped potential is in developing countries, for example in Africa. In 2008, 3,288 TWh of hydroelectricity was produced worldwide, which accounted for around 16 percent of total electricity.

Wind can also be used to spin the generator, but only when it is windy. Wind power is therefore independent of the wishes and needs of the electricity end-user and his or her needs. This makes wind power low in emissions but, given its intermittency, unreliable, especially at a local level. Spreading wind turbines to a larger area helps the intermittency somewhat, but it also increases the costs for the electricity grid.

As the proportion of wind power grows, so do its problems and its need for back-up power plants. The general consensus has been that if wind power stays under 20 percent of total electricity it is relatively easy to manage, but this depends on the structure of the electricity generation fleet. The best options to "load-follow" intermittent electricity sources such as wind and solar are hydropower, natural gas turbines and diesel generators. Nuclear has also been used in the load-following mode in Germany and France, for example.[123] Electricity with wind power has been very cheap from time to time, but this is largely because the utilities are dumping the electricity onto the market for free because it would be more costly to load-follow with their coal fleet. Renewable electricity also has primary access to the network in many places, which means that it always has to be bought first.

Since wind power requires almost its total capacity worth of back-up power, it can also be seen as a way to reduce fuel use in fossil power plants. This reduces the efficiency of these plants and makes them more uneconomical to use (but still mandatory). There have been plans, with Germany as the prime mover, to introduce a reserve-tariff system for plants that need to be on standby but operate at low capacity on average. This means that the end user gets to pay a tariff first on the solar or wind power produced, and then also pay a tariff for the power not produced by the back-up fossil power.[124]

The requirements for this load-following can be decreased by smart grids, intercontinental super grids, and overbuilding wind capacity (which leads to part of the peak production being abandoned). Storing energy in batteries, as pressurized air, as hydrogen mixed with natural gas, and as pumped hydro are some ways to alleviate the problems. They all add to the total costs, decrease capacity factors and efficiencies, have negative environmental effects, and produce emissions. Pumped hydro, where water is pumped up into the reservoir when there is surplus electricity, operates at 70–85 percent efficiency. The global pumped hydro capacity is 127 GW.[125]

The strengths of electricity

Electric motors have a far superior efficiency compared with internal combustion engines. A regular car with an internal combustion engine (ICE) can rightly be called a heating machine that has a side-effect of moving around. The efficiency of electric cars on the other hand is so good that heating the cabin becomes a problem in colder climates. Somewhat ironically, the solution presented has been a cabin heater that uses liquid fuels. Electric vehicles (EVs) also do not pollute locally, but they are far from emissions-free on a lifetime analysis. LowCVP did a study in the UK which concluded that half of the emissions from an EV are emitted in its construction, with batteries accounting for half of these. The average emissions of the electricity production also have an effect. Currently, the total emissions of an EV are maybe a quarter of those from a comparable regular vehicle.[126]

Electric rail (in its many forms[127]) is one of the most energy-efficient rapid modes of moving longer distances.[128] It remains as one of the central ways to lessen society's oil dependence. It also reduces air pollution and directs zoning to be less car-dependent.

The weaknesses of electricity

By far the biggest problem with electricity is that it is hard to store. Batteries are expensive compared with their capacity, power, and durability. Storing large amounts of electricity cheaply is one of the central research topics around the world. It is indicative that the lead-acid battery that can be found in every car is still one of the cheapest and most durable ways to store electricity. Additional limits are recharging and discharging speeds, safety considerations, and limits to size and weight.

Transmission losses are also a problem in some countries. In India it is not uncommon for 30 percent of electricity to disappear on its way. Electrifying vehicles and society is much harder in these kinds of circumstances. Transmission lines using HVDC (high voltage direct current) can move large amounts of electricity for longer distances, but they can't replace the regular AC power grid.

The biggest problems with EVs are the low power density and high price of the batteries. Since 2012, there have been an increasing number of EVs on the market. Their driving distance on one charge is usually below 200 km, and perhaps only 100 km with cold and demanding circumstances. This causes anxiety for many. When the battery alone represents maybe 10,000 euros in the price of the car, EVs are not very competitive. Many governments have seen it necessary to offer rebates to EV buyers. The manufacturing demands many rare earth minerals that have limited production capacity globally. This also puts limits on how many electric cars we can produce globally, no matter what their demand is. In theory, the global production capacity for lithium-ion batteries would be sucked up by just a few hundred thousand Tesla Model S electric cars produced per year. In real life, this is not even possible.[129]

The IEA estimated in WEO 2010 that the combined sales of EVs and hybrids will stay well below 10 million in 2020, but sales are to grow sixfold by 2035. Comparably, the world currently has a fleet of around 1 billion vehicles with internal combustion engines. California is investing 120 million dollars in a public EV charging network and expects 1.5 million zero-emissions vehicles on the roads by 2025. They would account for less than 5 percent of California's fleet.[130] In practice, every new technology is funded by the middle class. To put it in perspective, if a new EV costs 25,000 dollars, replacing the global fleet would cost 25 trillion dollars. The global GDP was 85 trillion (PPP) and 72 trillion (nominal prices) in 2012.

It is clear that every new hybrid or EV lessens oil dependency, makes the transition to a world of depleting oil production a little bit softer, and lessens the effects of any sudden oil crisis. The worrying aspect is how dependent many nations are on high taxes for gasoline and diesel. Is it possible, or even desirable, to lessen a nation's oil use in the short term, if every liter of gasoline brings half of its price as tax revenues to the state (in Europe at least)?

EV batteries have also been envisioned by many as enablers of a larger proportion of intermittent electricity production, as the batteries can act as storage. With current prices of maybe 10,000 dollars per 20 KWh of storage where electricity costs maybe 1 euro per 20 KWh, this makes little sense, especially since the batteries wear out if they are constantly charged and discharged. Still, Nissan went ahead in 2012 and brought its home battery station to the Japanese markets. It can charge the Nissan Leaf to 80 percent capacity in 4 hours, and it has its own battery that it can charge at night with cheaper electricity. It can also be used to power the home in case of a blackout.[131] The promising battery maker in the U.S., A123, filed for bankruptcy in late 2012, despite government funding.[132]

Hydrogen

Hydrogen is also one of those oil replacements that is not really feasible in any direct way. Further, it is not an energy source as such, but an energy carrier. There is little infrastructure ready for its use in transportation. This makes hydrogen a very long-term solution, decades away, if at all. There are a host of expensive problems yet to be solved with the hydrogen economy. These are rarely mentioned by hydrogen supporters.

Hydrogen is the most common element in the universe. Yet it is not found by itself in nature. It can be found in other elemental compounds (such as water, natural gas, oil, etc.). This means that pure hydrogen needs to be manufactured. This always takes energy, with eventual losses. Any society that plans to move from a hydrocarbon economy to a hydrogen economy will have to manufacture the hydrogen first with some other energy source.

Currently industrial hydrogen is mainly produced from natural gas (about 90 percent). This is a relatively simple process with an efficiency of around 80 percent. Somewhat ironically, this hydrogen is often used in the oil refining industry, for example to make thick bitumen more liquid to get it moving in the pipelines. The main use of hydrogen is as a feedstock to make ammonia, most of which is then used to make nitrogen fertilizers for the agricultural industry.

Hydrogen can also be produced from water with electrolysis. Around 5 percent of hydrogen is produced in this way. The efficiency, depending on source, is between 50 and 80 percent.[133] If the electricity is first made with fossil fuels in a power plant with an efficiency of 30–40 percent, this means that the total efficiency from primary energy to hydrogen is perhaps around 20 percent. A potential way to make clean hydrogen in the future could be to use surplus electricity from overbuilt renewables or nuclear power plants. This could happen at times when it would be more expensive to load-follow the electricity demand than to keep the plants humming. The process would also yield pure oxygen.

The third way to produce hydrogen is thermolysis. In this process heat causes the water molecules to become unstable. Thermolysis with organic matter is called pyrolysis. Without catalysts, thermolysis requires extreme temperatures of more than 2,000°C (3,600°F), which is hard to produce and upkeep. With catalysts the temperatures can be lowered. Some planned fourth-generation very high temperature reactor (VHTR) nuclear power plants can produce the 850–1000°C that is required for thermolysis with catalysts. This kind of VHTR would have the advantage that it could switch between producing electricity and hydrogen depending on electricity demand.

Concentrated solar power (CSP) is another option to produce the 1,000°C that is required. Hydrosol II, a 100 kW experimental CSP plant has been in use in Spain since 2008, and it is possible to scale the technology up to commercial scale.[134]

There are other ways to produce hydrogen. One interesting method is to use algae and sunlight in a bioreactor. Normally hydrogen is a by-product in algae photosynthesis, but by genetic manipulation the share of hydrogen produced can be increased. However, there are always risks. One is the proliferation of genetically modified algae from the required massive outdoor algae-pools. Another risk is the spread of regular algae into the production pools which would disrupt the production.

As always, the crucial factors between different processes are their efficiency, scalability and price. Iceland had fifteen hydrogen buses in the early 2000s. The goal was to move to a hydrogen economy by 2040. Since then, the financial crash has postponed these plans by at least a decade. The plans have also been criticized, since electrification of the transport system has been seen as a better solution.[135]

The strengths of hydrogen

Burning hydrogen in fuel cells produces no pollution or emissions, as the only by-product is pure water. The image of California's former governor

Arnold Schwarzenegger drinking water that came from a tailpipe is a powerful one, and easily used for marketing the concept.

As hydrogen is currently used by many industrial processes and oil refining, there is already a marketplace for it. Many potential alternative fuels require hydrogen. In some cases, as with Iceland with its large geothermal energy potential, the production of hydrogen is not a big problem. There are other issues however.

The weaknesses of hydrogen

Hydrogen is the smallest element. This makes storing and transporting it a tricky business. It has the habit of leaking from the best storage, and the current natural gas pipelines are not tight enough for pure hydrogen. There has been research to mix hydrogen with natural gas, which would allow current pipelines to be used for transportation and storage,[136 137] A pure hydrogen economy would need a totally new infrastructure. This would be slow and expensive to build as it would require massive amounts of energy and raw materials.

Uncompressed hydrogen also has a poor energy density. One kilogram of uncompressed (1 Bar, 20°C) hydrogen takes around 11 cubic meters (388 cubic feet). It has the energy content of 4 liters of gasoline. Compression alleviates this problem, but it takes energy. Compressed hydrogen is also even more likely to leak. Another way is to liquefy hydrogen by freezing it to –253°C (–423°F) but this also takes a lot of energy. Further, keeping the low temperature also takes energy. It is no surprise that the few hydrogen vehicles sold have astronomical prices. There are some plans to bring mass-produced hydrogen vehicles to the market. At the 2012 Paris car show Hyundai presented the IX35 Fuel Cell vehicle that is supposed to enter mass production. The first aim is to produce 1,000 of these by 2015.[138]

Hydrogen also has the nasty habit of combining with oxygen in an explosive manner. The image of the burning Hindenburg, which sealed the superiority of the airplane for air travel, comes to mind. Technology has progressed since Hindenburg's times, but bringing hydrogen to the public creates a number of safety hazards that need to be overcome, which adds to the cost. Fuel cells also need further development before commercial roll-out. Currently they use expensive materials and are often not very durable.

The worst weakness of hydrogen is its inherent inefficiency. Transferring electricity from the power plant to the end user usually has an efficiency of more than 90 percent. By comparison, the process of transferring electricity to the hydrogen manufacturing plant, making hydrogen with electrolysis, compressing/liquefying it, transporting it to the refilling station, and then burning it in a fuel cell has a total efficiency of maybe 20–30 percent.

Any significant move towards a hydrogen economy would call for huge amounts of clean surplus energy production capacity. If we first manage to replace current fossil energy with low emission power from renewables and nuclear, the use of hydrogen as a temporary storage and back-up power could be feasible. For this, there are potentially cheaper alternatives, such as pumped hydro, pressurized air, ammonia, butane, synthetic methane, and a number of other compounds that can be manufactured from hydrogen and are easier to store and transport.

Today, and in the near future, talking about a hydrogen economy seems more like avoiding the real subject than offering a real solution.

Notes

1 IEA (2012), World Energy Outlook Special Report 2012 - Iraq Energy Outlook, www.iea.org/publications/freepublications/publication/WEO_2012_Iraq_Energy_OutlookFINAL.pdf.
2 www.cia.gov/library/publications/the-world-factbook/geos/iz.html.
3 Gilbers, R., Perl, A. (2010), *Transport Revolutions*, New Society Publishers, p. 115.
4 www.youtube.com/watch?v=EIJ2AkebgyI&feature=player_detailpage#t=808s.
5 Gilbers, R., Perl, A. (2010), *Transport Revolutions*, New Society Publishers, p.1.
6 Tainter, J., Patzek, W. (2011), *Drilling Down*, Springer-Verlag New York Inc.
7 www.economist.com/node/15582301.
8 www.energybulletin.net/stories/2012-05-14/updating-world-deepwater-oil-gas-discovery.
9 http://seekingalpha.com/article/650701-brazil-and-petrobras-oil-potential-or-oil-dream.
10 http://brazilportal.wordpress.com/2012/05/11/how-brazil-is-making-an-example-of-chevron.
11 www.resilience.org/stories/2012-05-14/updating-world-deepwater-oil-gas-discovery.
12 www.huffingtonpost.com/2012/07/10/bp-liberty-project-alaska-ak_n_1660610.html.
13 Downey, M. (2009), *Oil 101*, Wooden Table Press LLC, p. 43.
14 Nikiforuk, A. (2010), *Tar Sands*, Greystone Books.
15 www.energy.alberta.ca/OilSands/791.asp#Production.
16 Downey, M. (2009), *Oil 101*, Wooden Table Press LLC, p. 45.
17 http://ostseis.anl.gov/guide/tarsands/index.cfm.
18 Aleklett, K. (2012), *Peeking at Peak Oil*, Springer, p. 101.
19 http://en.wikipedia.org/wiki/Caterpillar_797.
20 http://landofmachines.com/trucks/caterpillar-797.
21 www.energy.alberta.ca/OilSands/791.asp.
22 www.energy.alberta.ca/pdf/OSSResponsibleActionsImplementation.pdf.
23 http://environmentaldefence.ca/sites/default/files/report_files/TailingsReport_FinalDec8.pdf.
24 www.capp.ca/getdoc.aspx?DocId=193756&DT=NTV.
25 http://aleklett.wordpress.com/2013/07/18/rapporterade-oljefyndigheter-i-israel-saknar-betydelse-for-den-globala-oljeproduktionen.
26 www.eia.gov/analysis/studies/worldshalegas.
27 http://pubs.usgs.gov/fs/2013/3013/fs2013-3013.pdf.
28 www.businessinsider.com/california-shale-oil-reserves-2013-2.
29 www.latimes.com/business/la-fi-oil-20140521-story.html.
30 Hughes, D. (2013), Drill, baby, drill – Can unconventional fuels usher in a new era of energy abundance?, Post Carbon Institute, www.postcarbon.org/reports/DBD-report-FINAL.pdf.

31 www.theoildrum.com/node/9569.
32 www.eia.gov/todayinenergy/detail.cfm?id=15571.
33 www.theoildrum.com/node/9569.
34 http://businessdayonline.com/2014/03/decline-in-iocs-capex-amid-growing-oil-gas-discoveries.
35 www.forbes.com/sites/insead/2013/05/08/shale-oil-and-gas-the-contrarian-view.
36 Hughes, D. (2013), Drill, baby, drill – Can unconventional fuels usher in a new era of energy abudance?, Post Carbon Institute, www.postcarbon.org/reports/DBD-report-FINAL.pdf.
37 www.rosneft.com/attach/0/02/99/cera_week_en.pdf.
38 http://relooney.fatcow.com/0_New_12053.pdf.
39 www.nytimes.com/2012/11/14/business/energy-environment/russia-anticipates-boom-in-oil-extracted-from-shale.html?_r=0.
40 www.theoildrum.com/node/9569.
41 www.ft.com/intl/cms/s/0/45ffb86e-9146-11e2-b839-00144feabdc0.html#axzz2aRP95lR7.
42 http://nextbigfuture.com/2012/06/126-trillion-barrels-of-oil-equivalent.html.
43 www.forbes.com/sites/christopherhelman/2012/06/04/bakken-bazhenov-shale-oil.
44 http://aleklett.wordpress.com/2013/01/20/u-s-rig-count-for-oil-and-gas-and-future-economic-growth.
45 www.nytimes.com/2012/11/14/business/energy-environment/russia-anticipates-boom-in-oil-extracted-from-shale.html.
46 www.reuters.com/article/2013/03/18/russia-oil-tax-idUSL6N0C7FLT20130318.
47 Gustafson, T. (2012), *Wheel of Fortune: The Battle for Oil and Power in Russia*, Belknap Press.
48 Shale oil: The next energy revolution, www.pwc.com/en_GX/gx/oil-gas-energy/publications/pdfs/pwc-shale-oil.pdf.
49 http://ftalphaville.ft.com/blog/2012/08/13/1101721/shale-oil-everywhere-for-a-while.
50 www.forbes.com/sites/christopherhelman/2012/07/03/exxons-tillerson-speaks-some-convenient-truths.
51 Powers, B. (2013), *Cold, Hungry and in the Dark: Exploding the Natural Gas Supply Myth*, New Society Publishers.
52 www.bloomberg.com/news/2013-08-18/shale-grab-in-u-s-stalls-as-falling-values-repel-buyers.html.
53 www.zerohedge.com/news/2013-10-07/guest-post-about-shale-oil-gas-miracle.
54 http://theautomaticearth.com/Energy/the-darker-shades-of-shale.html.
55 http://shalebubble.org/wall-street.
56 www.theoildrum.com/node/9506.
57 http://boldnebraska.org/uploaded/pdf//HeadwatersInitiative-CondensateFactSheet.pdf.
58 http://world.time.com/2013/04/30/can-israel-become-as-oil-rich-as-saudi-arabia.
59 http://pubs.usgs.gov/fs/2012/3145/FS12-3145.pdf.
60 Legget, J. (2005), *Half Gone*, Portobello Books Ltd.
61 Downey, M. (2009), *Oil 101*, Wooden Table Press LLC, pp. 48–49.
62 Downey, M. (2009), *Oil 101*, Wooden Table Press LLC, pp. 48–49.
63 http://world.time.com/2013/04/30/can-israel-become-as-oil-rich-as-saudi-arabia.
64 www.westernresourceadvocates.org/land/oseroi.php.
65 Deffeyes, K. (2005), *Beyond Oil*, Hill and Wang, pp. 120–123.
66 http://en.wikipedia.org/wiki/Oil_shale#Environmental_considerations.
67 www.ymparisto.fi/download.asp?contentid=99476&lan=fi.
68 www.energy.eu/stats/energy-electricity-consumption.html.

69 www.kagirahsap.com/06-kuresel-climate/climate-change/2%20Europe's_Forest.pdf.
70 http://epp.eurostat.ec.europa.eu/cache/ITY_OFFPUB/KS-FK-12-001/EN/KS-FK-12-001-EN.PDF.
71 Rapier, R. (2012), *Power Plays: Energy Options in the Age of Peak Oil*, Apress.
72 http://domesticfuel.com/2013/07/31/commercial-scale-cellulosic-ethanol-is-real.
73 Hannula, I., Kurkela, E. (2013), Liquid transportation fuels via large-scale fluidised-bed gasification of lignocellulosic biomass, VTT Technical Research Centre of Finland.
74 www.catf.us/fossil/problems/power_plants/existing.
75 http://manhaz.cyf.gov.pl/manhaz/strona_konferencja_EAE-2001/15%20-%20Polenp~1.pdf.
76 Höök, M., Aleklett, K. (2009). A review on coal-to-liquid fuels and its coal consumption, *International Journal of Energy Research*, 34, 10, 848–864 (Wiley InterScience).
77 www.youtube.com/watch?v=ZYDPbl98q-Q&feature=player_detailpage#t=590s.
78 http://fabiusmaximus.wordpress.com/2011/02/14/24777.
79 www.businessday.co.za/articles/Content.aspx?id=133381.
80 Höök, M., Aleklett, K. (2009), A review on coal-to-liquid fuels and its coal consumption, *International Journal of Energy Research*, 34, 10, 848–864 (Wiley InterScience) p. 14.
81 www.eia.gov/energyexplained/index.cfm?page=coal_prices.
82 Höök, M., Aleklett, K. (2009), A review on coal-to-liquid fuels and its coal consumption, *International Journal of Energy Research*, 34, 10, 848–864 (Wiley InterScience).
83 www.treehugger.com/renewable-energy/jatropha-biofuel-yields-in-indian-waste lands-80-below-forecasts.html.
84 www.jouleunlimited.com.
85 http://en.wikipedia.org/wiki/Primary_production#Human_impact_and_appropriation.
86 Giampietro, M., Mayumi, K. (2009), *The Biofuel Delusion – The fallacy of large-scale argo-biofuel production*, Earthscan, p. 180.
87 Brown, L. (2012), *Full Planet, Empty Plates - The New Geopolitics of Food Scarcity*, W.W. Norton & Co.
88 www.triplepundit.com/2008/02/richard-bransons-new-push-biofuels-coconut-oil-fueled-airliner.
89 Aleklett, K. (2012), *Peeking at Peak Oil*, Springer, pp. 227–228.
90 IEA (2011), Medium Term Oil & Gas Markets, www.iea.org/Textbase/npsum/mtogm2011SUM.pdf.
91 www.guardian.co.uk/global-development/2012/sep/17/european-biofuel-targets-contributing-global-hunger.
92 Jylhä, P. (2011), Harvesting undelimbed Scots pine (*Pinus sylvestris* L.) from first thinnings for integrated production of kraft pulp and energy. *Dissertationes Forestales*, 133, p. 73 Available: www.metla.fi/dissertationes/df133.htm.
93 www.spiegel.de/international/world/world-bank-leak-biofuels-may-be-even-worse-than-first-thought-a-563927.html.
94 http://arxiv.org/pdf/1108.2455v1.pdf.
95 http://survivalacres.com/blog/we-are-now-one-year-away-from-global-food-riots.
96 *IEA World Energy Outlook 2011*.
97 www.oilcrisis.com/bentley/depletionOverview.pdf.
98 http://en.wikipedia.org/wiki/Peak_natural_gas.
99 *BP Statistical Review of World Energy 2013*.

100 http://epp.eurostat.ec.europa.eu/statistics_explained/index.php/Natural_gas_consumption_statistics.
101 Ratner, M., Belkin, P., Nichol, J., Woehrel, S. (2013), Europe's energy security: options and challenges to natural gas supply diversification, Congressional Research Service, www.fas.org/sgp/crs/row/R42405.pdf.
102 www.themoscowtimes.com/news/article/gazprom-to-remove-gas-price-discount-for-ukraine/495596.html.
103 http://theautomaticearth.com/Finance/shale-gas-reality-begins-to-dawn.html.
104 www.youtube.com/watch?v=1386Jt17myY&feature=player_detailpage#t=553s.
105 www.energysolutionsinc.com/naturalgas/Rig-Count-Statistics-29.htm.
106 www.theguardian.com/world/2012/aug/29/shtokman-russia-arctic-gas-shale.
107 www.slate.com/articles/health_and_science/future_tense/2011/12/is_there_really_100_years_worth_of_natural_gas_beneath_the_united_states_.html.
108 www.ted.com/talks/lang/en/t_boone_pickens_let_s_transform_energy_with_natural_gas.html.
109 www.bloomberg.com/news/2012-03-26/shale-boom-in-europe-fades-as-polish-wells-come-up-empty-energy.html.
110 www.hs.fi/talous/Liuskekaasun+pumppaus+alkoi+Puolassa/a1377840694279 (Link in Finnish).
111 Technically recoverable shale oil and shale gas resources: an assessment of 137 shale formations in 41 countries outside the United States, www.eia.gov/analysis/studies/worldshalegas.
112 http://en.wikipedia.org/wiki/Biogas.
113 www.gasum.fi/yritysinfo/media/uutiset/Documents/Panu%20Pasanen_Bionova.pdf (In Finnish).
114 Cuéllar, A., Webber, M. (2008), Cow power: the energy and emissions benefits of converting manure to biogas, *Environmental Research Letters,* 3, 034002, doi:10.1088/1748-9326/3/3/034002.
115 www.afdc.energy.gov/fuels/emerging_biogas.html.
116 2010 IEA World Energy Outlook.
117 Powers, B. (2013), *Cold, Hungry and in the Dark: Exploding the Natural Gas Supply Myth*, New Society Publishing.
118 www.csmonitor.com/Environment/Energy-Voices/2013/0704/A-surprising-source-of-demand-for-US-natural-gas.
119 http://fuelfix.com/blog/2013/03/14/natural-gas-exports-to-mexico-jump-24-percent.
120 www.youtube.com/watch?v=ZYDPbl98q-Q&feature=player_detailpage#t=590s.
121 http://cngeurope.com.
122 www.newscientist.com/article/dn7046-hydroelectric-powers-dirty-secret-revealed.html.
123 Lokhov, A. (2011), Load-following with nuclear power plants, *NEA News*, 29,2, www.oecd-nea.org/nea-news/2011/29-2/nea-news-29-2-load-following-e.pdf.
124 http://gastopowerjournal.com/regulationapolicy/item/897-germany-needs-over-2000-mw-of-winter-reserve-regulator.
125 http://en.wikipedia.org/wiki/Pumped-storage_hydroelectricity.
126 www.autogasforamerica.org/pdf/EV_fact_brief_11.25.pdf.
127 http://en.wikipedia.org/wiki/Hyperloop.
128 MacKay, D. (2008), *Sustainable Energy – Without the Hot Air*, UIT, www.withouthotair.com.
129 http://seekingalpha.com/article/1657582-teslas-crushing-battery-supply-constraints.

130 www.metroplugin.com/2012/04/california-commits-100-million-to-build-ev-charging-stations.
131 www.earthtechling.com/2012/06/nissan-leaf-home-power-back-up-system-rolls-out-in-japan.
132 www.nytimes.com/2012/10/17/business/battery-maker-a123-systems-files-for-bankruptcy.html.
133 http://en.wikipedia.org/wiki/Electrolysis_of_water#Efficiency.
134 http://en.wikipedia.org/wiki/Water_splitting.
135 www.csmonitor.com/Innovation/Energy/2009/0212/iceland-strides-toward-a-hydrogen-economy.
136 www.green-energy-news.com/arch/nrgs2011/20110085.html.
137 www.ogj.com/articles/print/vol-110/issue-4a/regular-features/journally-speaking/all-of-the-above-at.html.
138 www.autotoday.fi/page.php?page_id=55&news_id=201207813.

4 Oil in the modern, industrialized society

Oil as the fuel and lubricant of societies

Industrialized societies run on oil. Where we use this oil depends on the structure of the society, standard of living, and possible domestic oil production. Currently, the role of oil can be described as critical. Because of the historically vast volumes, energy content, and other unique properties of oil, it cannot be easily replaced with anything else.

Since the lessons of the 1970s' oil and energy crises, most developed countries use little oil for electricity production. The situation in the Middle East is totally different. Diesel-powered generators are cheap to install, with a fast return on investment that meets the needs of the rapidly growing demand for electricity. The actual fuel is often almost free due to government subsidies. Also, in many developed countries the reason for local diesel generators is that there simply is no central electricity network available. Even without subsidies, a local generator ensures the availability of electricity in the case that electricity distribution is not reliable.

The oil use of the world in 2012 was distributed to different sectors as follows:[1]

- 53.4 percent – Transportation
- 13.6 percent – Petrochemicals
- 5.8 percent – Other industry
- 8.7 percent – Buildings
- 6.3 percent – Power generation
- 12.1 percent – Other (agriculture, bitumen, lubricants and other non-energy use).

In OECD countries the share of transportation is higher than the global average, and the share of electricity production is lower. In developing countries, as much as 10 percent of oil is spent on electricity generation, while the share of transportation is around 50 percent.[2]

Around 75 percent of the share of transportation is used in road transportation, and roughly two-thirds of road transportation (45 percent of total transportation oil use) is used by personal vehicles. The rest is used on freight.[3]

The twelve biggest oil producers account for around two-thirds of all oil production. The three biggest, Saudi Arabia, the United States, and Russia, produce around one-third of the entire world's oil. For decades, the United States was also the biggest oil importer. In October 2013, the EIA reported that China had just surpassed the U.S. as the biggest oil importer. In September 2013 China had net imports of 6.3 mbpd, while the U.S. had net imports of 6.1 mbpd. There are three main reasons for the rise of China's imports:

1 Consumption in China is growing rapidly without much growth in its own oil production.
2 Oil production in the United States is growing.
3 The United States is reducing its oil consumption.

If the U.S. economy continues its recovery that started in 2013, it remains to be seen how that will affect its own oil consumption. The majority of reductions in oil use are likely to be due to economic problems. The growth in U.S. domestic production, largely due to shale oil, will likely hit its peak in the second half of this decade according to the IEA (WEO 2013) and other experts. The rate of China's economic growth (and therefore oil consumption) has shown a slowing trend in the last couple of years. Japan has increased its oil consumption in spite of its economic problems, as it has been replacing its nuclear fleet with fossil energy after the Fukushima accident. Saudi Arabia and Russia are by far the biggest oil exporters.

Table 4.1 Top 12 producers in 2012

Top 12 producers in 2012	*All liquids EIA/mbpd*	*Crude oil BP/mbpd*
Saudi Arabia	11.7	11.5
United States	11.1	8.9
Russian Federation	10.4	10.6
People's Rep. of China	4.4	4.2
Canada	3.9	3.7
Islamic Rep. of Iran	3.6	3.7
United Arab Emirates	3.2	3.4
Iraq	3.0	3.1
Mexico	2.9	2.9
Kuwait	2.8	3.1
Nigeria	2.5	2.4
Venezuela	2.5	2.7

Top twelve oil producers according to EIA (2013) and BP *Statistical Review of World Energy* 2013.

Table 4.2 Top 12 exporters in 2012

Top 12 exporters in 2012	*EIA/mbpd*
1 Saudi Arabia	8.9
2 Russia	7.2
3 United Arab Emirates	2.6
4 Kuwait	2.4
5 Nigeria	2.3
6 Iraq	2.2
7 Iran	1.9
8 Angola	1.8
9 Venezuela	1.7
10 Norway	1.7
11 Canada	1.6
12 Algeria	1.5

Top twelve oil exporters according to EIA (2013)

Table 4.3 Top 12 importers in 2012

Top 12 importers in 2012	*EIA/mbpd*
1 United States	7.4
2 China	5.9
3 Japan	4.6
4 India	2.6
5 Korea, South	2.2
6 Germany	2.2
7 France	1.7
8 Singapore	1.4
9 Spain	1.3
10 Italy	1.2
11 Taiwan	1.1
12 Netherlands	0.9

Top twelve oil importers according to EIA (2013)

Europe and oil

The oil production of the European Union is around 1.5 mbpd. It peaked in 1999 at 3.7 mbpd, and has since then declined by 60 percent. The EU consumed 12.77 mbpd in 2013. With current oil prices at around 110 US$ per barrel, the EU paid 1.2 billion US$ each day for its oil imports. With the current, prolonged economic problems, oil consumption especially within the eurozone has been dropping fast. Only a small part of this can be accounted for by better fuel efficiency of vehicles.

The figure above shows some of the historical turning points in their oil use for some EU countries. For many of the older industrial nations, the peak was in the 1970s. After this their consumption was reduced for some

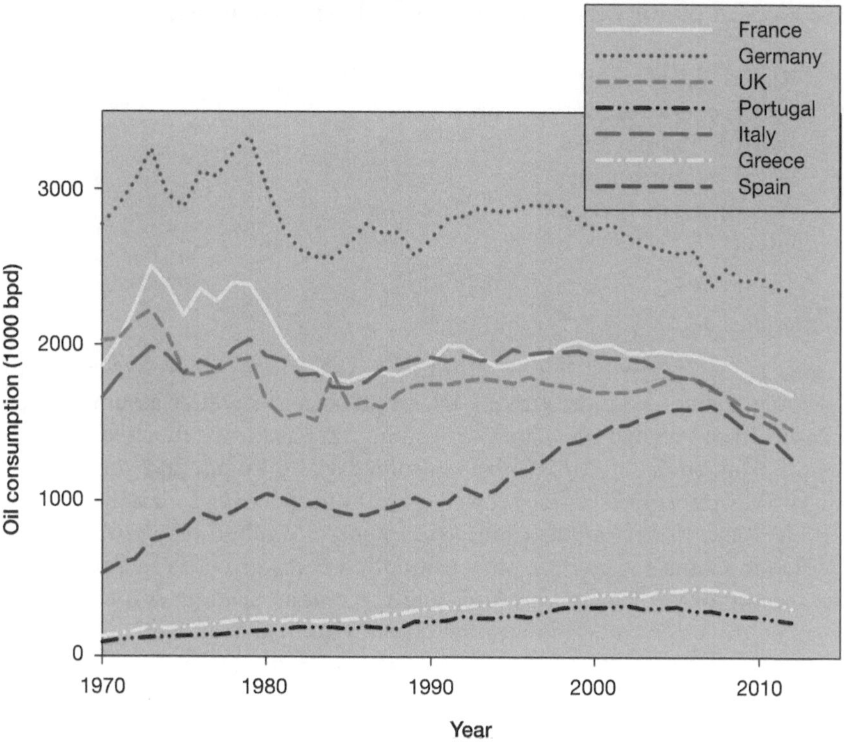

Figure 4.1 Trends in oil consumption of some European Union member countries, 1970–2013

Some of the countries suffering worst from the financial crisis have seen drastic reductions of around 30 percent in their oil consumption in just a decade or so.

Source: BP *Statistical Review of World Energy* 2014.

years, as electricity generation moved from oil to nuclear and coal due to the oil crises. The second peak for many countries came in the 2000s, after which the drop in oil consumption has been extremely steep for some countries. The reasons can be found first and foremost in the economic disaster of recent years.

But have the economic recession and related problems been due to reduction in oil use, or has the oil use dropped because of economic problems? Actually, this is a self-reinforcing feedback loop, where one factor reinforces the other. Many of the European countries have, for now, been shrouding this trend by going deeper into debt. As the European manufacturing industries have been decommissioned and moved overseas, so the amount of oil used and export revenues have been systemically reduced. The ability of EU nations to pay for their oil imports (and manufactured goods imports) has been eroding.

European oil imports come from the following countries:[4]

- Russia (33 percent)
- OPEC countries (35 percent)
- Norway (15 percent)
- Kazakhstan (5 percent)
- Azerbaijan (4 percent)
- Others (8 percent).

China and oil

China is the largest oil importer in the world, and its thirst for oil is still growing. China produces around 4.1 mbpd, and it is widely assumed to be near its peak production. China's consumption has grown much faster than its oil production. In 2013 China consumed 10.76 mbpd, and almost two-thirds was imported. China became a net importer over 2 decades ago, in 1993.[5] Predictions are that China will import 9.2 mbpd by 2020.[6]

Tied to China's growth in oil consumption is the growth of its economy. As the standard of living increases and the middle class grows, China has become the largest marketplace for cars. China has been willing and able to buy every drop of oil left over from the western nations wrestling with their economies. This has kept oil prices high. Oil consumption in China is only around 2.9 barrels of oil per person per year. By comparison, people living in the U.S. consume around 20 barrels, and Europeans consume around 10 barrels (depending on the country).

The United States and oil

Oil production in the United States peaked in early 1970s. It has been growing towards its second peak during the last few years, mostly due to tight oil from shale. The Energy Information Administration, EIA, estimates 2013 production to be around 7.5 mbpd, which is more than ever since 1989. For 2014, the EIA estimates growth to 8.4 mbpd.[7] This, with increases in fuel efficiency and the recent trend of driving less, has reduced the need for oil imports. The more enthusiastic pundits have started talking about the U.S. becoming oil independent and even a net exporter, considering all energy products. Optimism is of course needed to increase the confidence of consumers, which in turn helps the economy to grow. It is however irresponsible to create trust and confidence when they are not based on much reality. Consumer confidence will increase consumer debt, which will lead to worsening problems down the road.

In 2012, the United States imported around 7.4 mbpd.[8] In the first half of 2013, the net imports were around 6.5 mbpd. The five biggest importers were (by percentage of net imports):

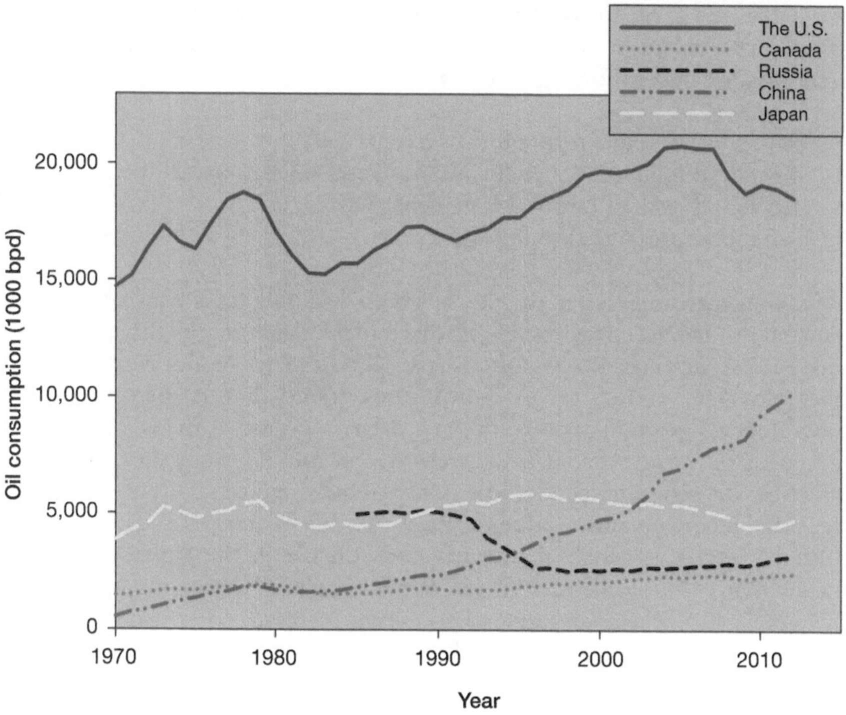

Figure 4.2 Historical oil consumption for the U.S., Canada, Russia, China, and Japan, 1970–2013

Russia has been and is still growing in the 2000s, and Japan is growing largely because it needs to replace lost electricity production from its nuclear fleet after Fukushima. China has seen a strong sustained growth.

Source: BP *Statistical Review of World Energy* 2014.

- Canada (34 percent)
- Saudi Arabia (18 percent)
- Venezuela (12 percent)
- Russia (10 percent)
- Mexico (6 percent).

U.S. citizens use about double the oil that the average European person uses. It is worth noting that their standard of living is not significantly higher than the standard of living in Europe. A large part of their consumption is such that it does not create much value to the consumer; the houses and the cars are bigger and the suburban sprawl around cities is less dense, but they are also less energy-efficient. This leads to transportation accounting for almost 70 percent of the U.S. oil use. From a global perspective this large percentage is something of an outlier. In contradiction to earlier estimates, the U.S. oil use has been declining since 2005. In early 2013, around 18.7 mbpd of oil products were refined in the United States.[9] These were divided into following categories:

- 9.3 mbpd gasoline (around 50 percent)
- 4.8 mbpd light distillates (including diesel oil, around 26 percent)
- 1.6 mbpd jet fuel (around 9 percent)
- 0.7 mbpd propane (around 4 percent).

Gasoline consumption in the U.S. has been declining for some years. About a quarter of the decline is due to increased fuel efficiency, and three-quarters due to less driving. People often drive less due to economic problems, unemployment and so on. A recent study found that 80 percent of the unwillingness of young people to get a driver's license in the U.S. is due to lack of a full-time job.[10] The increase in the fuel efficiency has had little effect on the situation as a whole. Only about 7 percent of the decline in the U.S. oil consumption is due to better mileage in vehicles.[11] There is also a kind of positive aspect to the frivolous oil use in the United States. It means that it should be easy to locate places where a dramatic increase in fuel efficiency is cheap and easy to do. These increases will often offer both economic and environmental benefits.

One is left with the feeling that it is important for U.S. citizens to be able to consume as much as they could ever want, where the Europeans have given the bureaucrats the tools to limit people's ability to consume (supposedly for the common good). The oil use in the U.S. affects the whole global community, and through high oil prices the global economy as well. It would serve a great many interests (if perhaps not the oil producers') if the general oil efficiency and consumption in the United States increased dramatically, closer to the European levels. This would lessen the pressure on the global oil prices.

Russia and oil

Like the United States, Russia has always been a major oil producer. In contrast to the U.S., Russia has also always been an important oil exporter. Russia has been through a lot of upheavals during the past few decades. The transition from the Soviet Union into the current centrally-controlled, bureaucratic and, according to many, corrupt democracy happened quite well or not so well, depending on who you ask. For example, the price of gasoline grew fivefold in one night, as the system collapsed.[12]

Currently, the Russian people are getting older fast, and the population is threatening to collapse. This is mainly due to low fertility and immigration rates and low life expectancy in Russia. The UN Development Program estimates that by 2025 the Russian population could shrink by another 10 percent from its current level of 143 million, and the trend continues downwards. Furthermore, Russia is not likely to have a large, young, unemployed and restless population that would have enough power to demand change in how the government is run. In this respect, the situation in Russia is totally different from many African and Middle Eastern nations.

The recent history of Russia's oil production is so interesting and complex that it deserves its own book. Thane Gustafson offers an interesting, if lengthy, overview of just that in his book *Wheel of Fortune: The Battle for Oil and Power in Russia* (2012). After the Soviet system collapsed, the oil sector that had been under tight state control quickly ended up in deep trouble. With the ensuing chaos, asset-grabbing and infrastructural collapse, oil production also collapsed. Since those days, much has happened. From the ruins of the Soviet oil sector rose the current generation of vertically integrated oil companies that explore, produce, refine and sell their product. In the 1990s, there was a privatization boom among the oil corporations. Those were also the years that the generation of the current oligarchs was born. As Putin took power, the state proceeded once again to nationalize its strategic oil and gas resources and assets. There were some oligarchs who disagreed. Mikhail Khodorkovsky, the wealthiest man in Russia in 2004 and owner of the oil company Yukos, was a prime example. They soon realized that their wealth that had been amassed in record time was disappearing even faster. This re-transfer of wealth and assets gave birth to another generation of oligarchs. Khodorkovsky faced prison, with a sentence lasting until 2017. Putin decided to pardon him and, just before Christmas 2013, he was released.[13] Yukos finally ended up in Rosneft's control. BP left Russia and its joint venture with Rosneft at Sakhalin-5 in late 2012. BP mentioned the high prices of developing the oil and gas resources, however Rosneft is still rather interested in the project.[14] Speculation has been that BP had worsening problems with the local legislation system.

In the IEA *World Energy Outlook* for 2011, Russia got its own section. The report seems to put a lot at stake on the assumption that Russia will manage to increase the energy efficiency of its society and economy to a comparable level with other OECD countries. If this happens, there will be more energy and natural resources left for export, and everybody would benefit. The report fails to account for the rebound effect, which could mean that even if Russian efficiency rises, its economy would grow that much faster (because it would be more efficient) and it could still end up using a similar amount of resources to what it would have used anyway. Also, it is wise to look at the possible reasons why Russia's ability to increase its efficiency has not already happened, and maybe won't happen in the future on a large enough scale.

1 The trend is against this. During the 2000s, Russia has been using up the Soviet-built infrastructure without investing nearly enough to improve it, despite the massive revenues from oil and gas. The requirements for the rotting infrastructure are so vast that even with

brisk economic growth it is unsure if there will be enough to even stop the current decay, let alone bring a swift change for the better.

2 The central government and the corrupt local bureaucracy will not necessarily have enough incentive to start massive reforms, since the current income is sufficient to keep the rich elite rich and the poorer people sufficiently satisfied. When this income starts to diminish in time, it will be too late to do much about it. In addition, Russia is not the most attractive place for foreign investment. In the 2011 World Bank Doing Business report, it was ranked 123 in building permits (out of 183 countries) and 162 in foreign trade (out of 182 countries). It will be very hard to attract enough investments from foreign sources, as the need and potential is huge.

A risky investment environment can be converted into numbers through an increase in the expected profit from an investment. For example, if a multinational oil company invests in a new project, as ExxonMobil is planning to do in Bazhenov, it requires profit according to perceived investment risk. In Canada, where the politics are stable and there is not much corruption, this profit margin can be around 8 percent after taxes. In Russia, the required margin can be near 20 percent due to added risks. This means that only the best projects attract investments, and the rest stay undeveloped, waiting for rising oil prices or a better investment environment.

Russia's oil reserves and production

The estimates of Russia's oil reserves (proven, or P1) vary depending on the source.[15]

For 2P (probable) reserves that have a 50 percent chance of being produced, including the likely oil and gas resources that can be found in the Arctic region, the best estimate might be around 100–150 billion barrels. Compared with the IEA estimates on future oil production, this is far too little. The IEA estimates that stable production of around 10 mbpd will continue up to 2035, which means around 87 billion barrels of aggregate production. This would assume a total reserve of at least 200 billion barrels,

Table 4.4 Russian oil reserves by different sources

CIA World Factbook (2013)	80 billion barrels
Oil & Gas Journal (2006)	60 billion barrels
World Oil (2006)	69 billion barrels
British Petroleum/BP (2014)	93 billion barrels
United States Geological Survey (2006)	116 billion barrels
Wood Mackenzie (2006)	120 billion barrels
Russia's official statement (2013)	130 billion barrels + 70 Gb 2P-reserves

which is quite high compared with most estimates. Of course with technological advancement and higher prices the reserves increase. The problem is that we probably can't afford to pay high enough prices.

Professor Kjell Aleklett casts some light on the Russian oil reserves in his book *Peeking at Peak Oil* (2012). He sees a realistic reserve estimate at around 120 billion barrels, which would place peak production around the middle of the current decade. This means that Russia will export over 6 mbpd during the next 10 years, if there are no bigger disruptions. After this, production will start to decline, and by 2035 Russia's exports will be below 2 mbpd. It remains to be seen how tight shale oil from Bazhenov will affect this estimate. There are plans to start production from Bazhenov during the current decade, and the resource base has been estimated to be significant.

Currently (2013) Russia exports around 7.4 mbpd, and consumes around 3.3 mbpd. When comparing energy content, Russia's natural gas production is similar in size to its oil production. The net exports of oil, along with total fossil fuel production, have been stagnant since 2007. After the Soviet system collapsed in the 1990s, Russia's oil production plunged to 6 mbpd with 4 mbpd of net exports for several years. Production started to recover in the early 2000s, along with the rising global oil prices. Money started flowing to the oil companies once more. Soon they found out that the central government was more than willing to direct much of that money to the state coffers. Various tax revenues from oil grew from 5.6 billion dollars in 1999 to a whopping 83 billion dollars by 2005. At the peak in 2008, the state took 90 percent of the profits.

Around two-thirds of Russia's oil production comes from Western Siberia. The production there is mature, and it requires constantly growing investments to keep steady. A lot of the fields were badly damaged during the Soviet system, as the central government kept increasing production quotas without regard for how they would be met. Repairing this damage, if indeed possible, will be expensive and require a high level of expertise not easily available. Despite investment, many of the big fields are declining. For example the expensive Multi-Zone Hydraulic Fracturing (MZHF) method has been used on Russian fields for years to slow down their decline.

Another important region, accounting for about a fifth of total production, is the Ural-Volga area. Back in the 1970s it was the most important production area of the Soviet Union. However, as its most important supergiant Romashkinskoye field has matured, the area's total production has been declining.

As the current areas mature and decline, the fields in East Siberia will have to play a bigger role in Russia's oil production. The main problem has been the lack of infrastructure. The first phase of the ESPO pipeline was completed in 2009. At 2,757 km (1,713 miles) in length, it reaches the export terminal of Skovorodino near the Chinese border. The second phase of ESPO was completed in late 2012. It grew the capacity from the previous

600,000 bpd to around 1 million bpd. It also extended the pipeline to reach the Koz'mino export terminal on the Pacific shore. Exports are divided between the U.S. (35 percent), Japan (30 percent) and China (28–36 percent).[16]

The Rosneft-operated Vankor field that is located north of the Arctic Circle is Russia's biggest oil find in 25 years. Since it started producing in 2009, it has been the main driver of any growth in Russia's total production capacity. Vankor currently produces over 300,000 bpd. Most of the other Russian Arctic production is natural gas. In the east, oil and gas are being developed on Sakhalin Island. Many of the fields are difficult and expensive to develop, for various reasons. A good example is the recently cancelled development of the Shtokman gas field, which had estimated development costs of around 40 billion dollars.

Russian oil production is controlled by state-owned corporations. The biggest producers are:

- 2.5 mbpd – Rosneft
- 1.8 mbpd – Lukoil
- 1.4 mbpd – TNK-BP (acquired by Rosneft in 2013)
- 1.2 mbpd – Surgutneftegaz
- 0.6 mbpd – Gazprom Neft
- 0.5 mbpd – Tatneft.

Rosneft is the undisputed king of the Russian oil hill. It is one of the largest energy companies in the world. It is also the prime choice when foreign companies look for joint venture opportunities or partnerships in Russian oil. Many foreign companies have had trouble in Russia. BP ended up selling its 50 percent share of TNK-BP to Rosneft after 10 years. ConocoPhillips also sold its 20 percent share of Lukoil.[17]

The majority of Russia's exports go to Europe. The largest export countries are:

- around 0.7 mbpd – Germany
- around 0.55 mbpd – the Netherlands
- around 0.42 mbpd – Poland
- around 0.38 mbpd – China
- around 0.2 mbpd – the U.S., Finland, France, and Sweden (each).

The state-owned Transneft handles 90 percent of oil transportation in Russia.

The oil trends in OECD countries

What do the recent oil trends mean for OECD nations? It means that there won't have to be any major crises or disturbances in the oil markets. The mere continuation of past trends will mean a steadily worsening situation for most OECD nations during the coming decade. Another way to put this is that *it is not possible for these trends to continue much longer*. The important question is what will break the trends, and what consequences this will have.

If the trends seen from 2005 to 2010 continue for the next decade, the amount of oil exports available to OECD countries will decrease by 40 percent, if production holds steady.[18] According to IMF economists, increasing oil production will be possible only if the price of oil has room to rise significantly. In the end it comes down to how much humanity is able to pay for oil without going into recession. If oil prices cannot rise from the current level, oil production cannot rise much either. Where can the OECD nations cut this much consumption without a serious recession or a steep decline in their standard of living?

The easiest way would be to shift electricity production from oil to something else. This has already been done as a result of the 1970s' oil crisis. Oil used for heating buildings is another that can relatively easily be converted to another energy source, but its share is not that big in most countries. Industrial heat can move from oil to coal or natural gas (or even electricity or nuclear waste heat, in some cases), but usually this is feasible only when the current equipment and facilities are near the end of their usable life. Cutting our use of oil in raw materials and lubricants is not easy either, as oil is an essential and cheap feedstock for many plastics and other goods. Increasing the use of biofuels means increased use of oil in agriculture and forestry. None of these sectors offer much towards cheap and easy cuts in our oil use.

We are left with transportation, which uses more than half of all oil. According to the trends, oil use in transportation must halve in a decade, or drop by 5 percent annually for years. Improving the fuel efficiency of the global car fleet can show a drop of a few percentage points. As a relatively small, yet still important contribution, freight can be transferred to more fuel-efficient transportation such as railways and river barges. The sales of electric vehicles and other alternatives are marginal compared with sales of gasoline and diesel-powered vehicles, especially as these cars, trucks, and buses will be in use for at least 10–15 years. There are no readily available alternative fuel configurations (other than natural gas vehicles) for heavy vehicles. There will also be pressure for bus transportation to grow if personal vehicles are used less. The development of technology is therefore essential, but it is not enough by itself to ensure positive development. In Europe a major obstacle is the slow and bureaucratic legislation system. This is confounded by resistance offered by some member countries and various interest groups. Car manufacturers tend to maximize their market advantages and lobby efficiently against fuel efficiency improvements when it suits them.

By lowering the highway speed limits, a few percent can be saved. The rest of the reductions must be made by transporting fewer goods and people. The majority of this will come from reducing the use of personal cars. What would it be like, for example, if you had to cut your current driving by a quarter? And how does this trend compare with the estimates that driving will actually increase, as is often seen in reports when new highways are planned?

It does not seem likely that these kinds of reductions can be made in a controlled and voluntary way, even if western youth are increasingly opting not to get a driver's license. Also, the sales of new cars in Europe are at their lowest level for 17 years. This means that driving will likely decrease in Europe over the long term, but on the other hand the older car fleet will be less fuel-efficient. It will also be interesting to see what will happen to consumption and business built on car travel, especially outside cities. Hypermarkets, hobbies, jobs, and most other services are currently quite dependent on the personal car.

It seems quite clear from the trend of decreasing oil use, shown above, that the current amount or mode of transportation cannot continue; yet another trend to break in this decade. The consequences of this are likely to be immense.

A sudden but prolonged crisis

If there was a sudden crisis with oil, what could we do? Short-term relief would be available from the emergency supplies that at least all IEA members (mostly OECD nations) are required to have for a few months. Oil companies and refineries also have their own oil storages, and armies have theirs. But what if the crisis was longer, and was due to a more permanent supply limit? Or what if the price of oil keeps getting higher and drives OECD economies deep into recession? How big a share of our transportation fuels could we substitute with synthetic fuels or biofuels? Moving to natural gas has increased the risk because there is not sufficient available emergency storage in Europe.

According to the experts interviewed, in less than 5 years practically nothing could be done. In our current societies, the planning and permits alone can take years. In a business-as-usual situation, a bigger advanced biorefinery or CtL plant can take 10 years from planning to operations, and the costs are in the hundreds of millions. One such plant can produce between 200,000 and 300,000 tons of biofuels annually. This corresponds to 4,000–6,000 barrels per day

With a longer time frame of 10-plus years, governments can divert available oil to ensure energy production and build alternative fuel plants and refineries. This would likely be accomplished under some form of emergency or martial law, which would allow the state and military to control the flow of goods and services instead of the current market-based method. This would also allow bypassing environmental legislation and regulation, therefore bypassing normal democracy, which could lead to civil unrest.

If oil imports in Europe decreased significantly, so would the living standards of the European people. Business in less critical sectors would quickly fall to a minimum, and probably the ensuing chaos would result in many more beneficial companies and organizations ending up in the bin. The shrinking purchasing power of people would be directed towards essential goods: food, lodging, and essential transportation and energy. Discretionary spending would crash. Many people would find out they were living a little too spaciously, expensively, and far away from work, that is, if they even still had a job. This would crash real estate values in certain areas with horrific consequences for the mortgage and banking businesses. All of these factors would lead to a deep recession and huge unemployment. This might be met by a transition to martial law.

Summary

The oil consumption of several countries that are still growing such as China, India, and a few others are likely to increase in relation to developed countries, such as those in Europe. Due to the economic stimulus provided by shale oil and gas, the trend of a decrease in oil use in the U.S. may also reverse for a few years. This reversal might be held back by the fact that the economic growth mainly concentrates on a relatively small minority of the population, the increase in fuel efficiency, and the historically expensive gasoline price (which is partly due to the fact that it is illegal to export oil but legal to export gasoline from the U.S. and so the gasoline prices have followed the international markets more readily).

It is worth noting that if we remove North America from the picture, the rest of the globe has not seen growth in oil production in recent years. And the growth in North America has been almost exclusively from unconventional oil. Conventional, and therefore much cheaper, production is still mainly located in the Middle East, so it is likely that their weight in a geopolitical sense will not decrease in the longer run. On the other hand, a rapid population explosion with a rapid rise in living standards will mean less oil being released to global markets. Also, a young, unemployed population, accelerating climate change, and inequality increase the risks for social unrest, which can greatly affect the ability of these countries to export oil.

Transportation and logistics

> In the short term, peak oil is not an energy crisis. It is a liquid fuel crisis.
>
> Robert Hirsch, 2011

Transportation is by far the biggest single oil-using sector, as more than half of global oil is burned to transport people and goods. Liquid fuels have many properties that make them ideal to fuel transportation. These include high energy density, and ease of transportation and refining. Oil was the crucial factor in the otherwise promising start of the electric car ending up in the bin a hundred years ago. Another accomplishment of cheap oil is the rapid globalization that started a few decades ago. Oil made the long-distance transportation of goods, people and raw materials not only possible, but cheap and fast as well. As the end of cheap oil begins in earnest, we may find that the hyper-globalization we have experienced is just a relatively short bump of a few decades in humanity's long history.

People and goods in the modern world have the need to move. The current just-in-time society is not possible without, for example, trucking services. Since most of our transportation is dependent on oil, our whole economy is more or less dependent on oil. Even our energy production is dependent on oil, as there will often be no energy to produce and consume unless we first burn oil to transport the fuels and other necessities. This impacts most of our economic activities. Almost nothing will happen in a society without the constant movement of goods and people. With the current infrastructure this transport is powered by oil.

The needed change is not just around the corner. For example in Europe there are plans to reduce trucking fuel use by 7–12 percent through more aerodynamic and longer trucks, but the legislation will take at least a couple of years. It will then take 3–5 years for the manufacturers to implement the needed changes, and then a further 10 years for the fleet to change. In addition, some manufacturers are actively lobbying against the emissions reductions targets in cars, which may cost another few years.

The personal car is a symbol of freedom, since it enables one to leave whenever and go almost wherever one pleases, with only the road and gas station network setting any limits. The speed is relatively high also, but the efficiency of the internal combustion engine remains rather mediocre, usually only 20–30 percent, of which only a fraction is used to actually transport the passengers. Railways are dependent on set schedules and the railway network, but trains are otherwise relatively fast and almost unbeatable in their energy efficiency. Freight use is limited by the fact that the freight usually needs to be moved to rubber wheels at some point anyway. This would demand time, space, and personnel, which all add to costs. Flying is by far the fastest way to move long distances, whereas bicycling is one of the most energy efficient ways to move, but is limited in terms of speed. In his book *Sustainable Energy – Without the Hot Air* (2008) David MacKay presents the energy efficiencies of most transportation methods.[19] The book is also available free of charge at www.withouthotair.com.[20]

An increase in personal vehicles and driving is a direct result of cheap oil. This has led to enabling and promoting it in society by building more roads, gas stations, and zoning on the premise of the passenger car. In central and northern Europe, this has been slowed down by expensive gasoline due to high taxes, and especially the structure of old cities and towns, high population density, and short traveling distances. Europe has also developed its railway network more than for example the United States.

The short history of the passenger car

> In the case of road transport, oil-based fuels still account for around 95 percent of total energy use – a share that is barely lower than in 1971 – due to practical and economic barriers to the deployment of alternative fuels.
>
> IEA *World Energy Outlook* 2013

In the early 1900s, the motorized carriage arrived to save the bigger cities from their growing pollution problem: that of horse manure. Many towns were quite literally drowning in the manure of the then ubiquitous method of transportation, the horse. With mass manufacturing and the relatively cheap Ford Model-T (sold from 1908 to 1927), driving started to increase and the horsewhip industry was driven to the margin. As the years passed, all functions of society started to be seen from the point of view of the personal car and the mobility it offered. This less efficient, but more flexible, mode of transportation started to take over from light rail, especially in the United States. The whole idea of the highway network was copied from the German autobahns, and it was the biggest construction project at the time in the United States. It was, as was the internet some decades later, built with military applications in mind. After decades of neglect, it is currently in rather bad shape with little money available for basic repairs. Only acute repairs are made.[21]

After the Second World War, another construction project was started in the United States, which some critics claim to be the worst investment in human history: suburban sprawl. These less dense areas were built on the assumption of free mobility and cheap fuel. It remains to be seen how useful or viable they are in a time of expensive and scarce oil. Zoning was done in a way that separated commercial areas, working places, and residential areas from each other. This means that even basic functions in a society demand a lot of moving around. As public mass transit systems have also almost been shut down, the people living in the suburbs face little choice: one must have a car and one must buy gasoline for it in order to go to work and buy groceries. The result has been that vast areas of fertile land have been covered with asphalt so that people can live as energy-inefficiently as possible.

Roads and vehicles are huge national investments, which cannot just be changed for another when we feel like it. We are married to them, if not forever after, at least for several decades, and their upkeep demands a lot of resources. One of the most harmful properties of the car is that it enables very energy-inefficient living in the suburban and exurban areas. Building infrastructure with many overlaps is usually not smart, as it will lead to less than optimal rates of use of both, with increasing costs for upkeep. In a world of limited and scarce resources, it is important to maximize the utilization rate for investments. Stranded assets in the form of infrastructure, vehicles and machines need to be avoided. Instead of building utopian visions, we need to design the future from what we now have, to be as resilient and flexible as possible, with a good grasp of the bigger picture.

Roads and transportation infrastructure

The European Union Road Federation (ERF) publishes a handbook about European road statistics, with the latest version from 2011, and statistics up to 2009.[22] According to this, transportation and storage were a business of over 1,200 billion euros. The turnover was divided as follows (2008):

- warehousing and support activities 36.6 percent
- road freight transport 26.7 percent
- air transport 10.9 percent
- road passenger transport (including short-distance mass transit) 9.1 percent
- sea transport 9 percent
- railways 6 percent
- pipelines 1 percent
- inland water transport 0.7 percent.

The aim in Europe is to increase the size of shipments. This increases pressure to grow available temporary warehousing capacity, as goods will move in bigger shipments but less often. This means that costs are being moved from transportation to warehousing. The same will result if we reduce the speed of transportation. The need for warehousing will grow, since all the participants are dependent on fast, just-in-time deliveries.

If the volumes remain the same, a smaller turnover in transportation means increased efficiency for the underlying society. Concentrating too much on turnover makes it easy to forget that it tells us only the price of the transportation services we are getting. In ton-kilometers the respective sectors accounted for the following volumes and shares (2009):

- road transport 1,691 billion tkm (46.6 percent)
- sea transport 1,336 billion tkm (36.8 percent)
- railways 362 billion tkm (10 percent)
- inland water transport 120 billion tkm (3.3 percent)
- pipelines 120 billion tkm (3.3 percent)
- air transport 2.5 billion tkm (0.1 percent).

Economic recession shows clearly in transportation statistics. Between 2007 and 2009, European transportation decreased over 15 percent to a level similar to the start of the century. In 2012 it decreased by 4 percent from the previous year. The combined transportation in 2009 was 3,632 billion tkm, while the peak in 2007 was 4,175 billion tkm.

Recession cuts our oil consumption, but declining oil use also cuts our real economy. In other words, we could say that, holding all other factors constant, a smaller amount of oil can be used to run a smaller economy. The EU27 countries had over 9 million people working in the transportation sector, with 5 million in road transportation. Of these, 3 million were in freight and 2 million in passenger transport. Investments in transportation infrastructure are normally around 1 percent of GDP, depending on the country. In Western Europe, around two-thirds of the investment goes into roads, and nearly one-third is spent on railways. In Eastern Europe, more than four-fifths are spent on roads, with only 13 percent spent on railways.

The road infrastructure is being eroded in many countries. In Germany it costs 2.5 billion euros annually to keep the road network in good shape. Since the turn of the century, 25–30 billion of "debt" has accumulated. Many bridges over the Rhine have restrictions on heavy vehicles. Spain has relatively the largest highway network in the world, but there does not seem to be enough money for the upkeep. Historically, the EU has had the best road network in the world, but this might be changing as China increases funding for road network investments. The EU has also looked at the situation through the perspective of traffic congestion. On average, around 1 percent of GDP is wasted by sitting in traffic jams, with higher portions of 2–3 percent in some countries such as the Benelux countries.

Rubber wheels today

Almost everything that moves on rubber wheels uses an internal combustion engine. There are roughly 1 billion vehicles in the world, and almost all of them have an ICE. Both the U.S. and China, the biggest car markets in the world, have around 250 million passenger cars each.

In Europe, registrations of new cars have been falling since 2007. In 2010 there were over 13 million new registrations, while in 2007 there were over

Table 4.5 Paved roads and railways in some countries and areas (2008)

	Paved roads 1000 km	*Meters per inhabitant*	*Meters per km²*	*Railway network 1,000 km*	*Meters per inhabitant*	*Meters per km²*
EU-27	5,000	10	1,130	212.9	0.42	48
U.S.	4,400	14	574	200	0.63	26
Japan	962	8	2,639	273	2.13	749
China	2,779	2	293	79.9	0.06	8
Russia	754	5	44	86	0.60	5

15 million. More than one-fifth of these were in Germany. France, Italy, and the UK had around 15 percent of EU registrations each. Thanks to directives, the emissions of new cars have been falling in the 2000s. German car manufacturers BMW and Daimler-Benz have been lobbying hard for a loosening of the emissions directives.[23] In 2001 the average emissions of a new car were 169 g/km. In 2010 the same average was 143 g/km, although the EU has been criticizing the car manufacturers for making the emissions of new cars seem less than they actually are.[24] Making the engine smaller has a strong correlation with also making the emissions smaller, although the engine power and vehicle mass have also been declining slightly.[25] In 2013, around 15.7 million new vehicles were sold in the United States.[26] Over half of these were light passenger cars with the rest being light trucks.

In addition to the popularity of light trucks in the U.S., even passenger cars are much bigger than their European counterparts. On average, they have 300 kg more mass, and light trucks have 500 kg more mass than in Europe. The size of the engine is 1 liter bigger in the U.S. (2.6 l vs. 1.6 l) and engine power is almost double. Europeans drive with much smaller vehicles, as do the Chinese. Over 60 percent of new cars sold in Europe belong to the three smallest categories, meaning they are smaller than average (think a VW Golf or smaller). The biggest and most powerful (and least fuel-efficient) cars in Europe are sold in Germany.

In Europe more than half of cars run on diesel oil. In the U.S. around 95 percent run on gasoline. Europeans rely on manual transmission (86 percent) while the U.S. citizens prefer automatic (around 80 percent).

China has been the world's largest market for new cars for some years. More than 18 million cars are manufactured there each year. This also resulted in China becoming the world's largest oil importer in late 2013. The Chinese government has implemented ever-tightening fuel efficiency standards. The goal is to drop from the 2010 level of 7.8 l/100 km to 6.9 l/100 km by 2015 for new cars. By 2020 new cars should be at 5 l/100km. In addition to lessening demand for oil imports and decreasing pollution and emissions, the Chinese car manufacturers aim to export their cars to foreign markets.[27]

Even if fuel efficiency keeps on improving, it does not mean that oil use in transportation will decrease, if the distance driven increases. The IEA published two road maps in autumn of 2012. According to these we could improve fuel efficiency by half by 2050, but at the same time, the number of vehicles would grow and the amount of oil burned in transportation would remain at current levels. In relation to the oil production trends presented in this book, the conflict is clear.[28]

The economists' answer to increasing fuel prices is often people driving less. From the viewpoint of a single person, the change might be all but gradual. As long as people have jobs and hobbies they can afford, with a regular salary, they will be likely to drive as much as before. The money going on more expensive gasoline is taken from somewhere else. When an individual loses their job due to the recession caused by the expensive oil, they also lose most of the reasons to drive. They also lose any realistic chance of buying a new hybrid, or any new car that would save them some money on gasoline bills. From the macroeconomic point of view, the decrease in total distance driven is shown as gradual, but for the individual, the change can be crushing and sudden.

The car

The passenger car is one of the central symbols of a modern, middle-class society and those belonging, or wanting to belong, to it. The freedom of movement it offers now, especially for families, has never previously been experienced to this extent by the human race. The relationship between humans of the industrial era and the internal combustion engine has been close and tight, since it has been one of the key drivers of our current standard (and way) of living. Even those who have no car enjoy the benefits indirectly, as the society around them, and the services it enables and offers, is made possible by the car.

On the other hand, the direct and indirect drawbacks of driving are spread everywhere. Local air pollution, noise, traffic accidents, congestion, traffic control and policing, the road network, and the living space that has to be reserved for the automobile are there for everyone to suffer, even if one does not own a car. Those without a car still usually end up paying for parking spaces and parking garages, directly or indirectly. Areas reserved for roads are away from other uses, whatever those uses might be. Zoning of neighborhoods and city districts is distorted as it is done from the perspective of the passenger car. The pollution and small particles kill and cause health problems for millions of humans annually.

The personal car does not make much sense when seen from an energy efficiency point of view, either. In effect, we dress up in a 2-ton suit of armor of steel and plastic, which we then use to move around on a vast network of paved roads and other infrastructure. Only a small percent of the energy is used to actually move the driver and passengers, as most of the energy content is wasted as heat in the engine. The current way we use energy with our habit of driving around is a prime example of waste that no species should be able to do from an evolutionary point of view. Humans have been able to afford it by tapping into the fossil fuels that were formed by millions of years of sunlight and geological processes. These fuels can only be used once, and the reckless waste we are participating in could well be a source of much wonder for some future generations.

The efficiency of the car can be increased in many ways. Many of these have a trade-off. We can move more slowly, with less freedom and comfort, but with less energy. With ride-sharing it takes longer, if we need to pick up or drop off people at different places along the way. With less speed, the air resistance and fuel consumption decreases, but it takes a longer time to get to the destination. Car sharing or renting means that the car is not immediately and always available, if the need arises, and so forth. In the current hyperactive and competitive world many seem to lack the opportunity to use more time and less energy for transportation.

The fuel efficiency of the car itself is also improving. Lighter materials, smaller cars, better engines, hybrids, and full electric vehicles are becoming more common every day. The fleet of cars takes a long time to change, and new technologies are introduced relatively seldom to the mass market. As a result, oil has continued its dominance for the time being. There are an array of substitute fuels that are easier and more efficient to make than gasoline. These include for example methanol and DME (dimethyl ether), but the problem is that they are not drop-in fuels for the current vehicles. One interesting possibility is the boron car. Using boron as a fuel would bring a closed fuel loop, as it is fully recyclable to be used again, even if the recycling naturally uses energy. Each boron car would require one tank full of boron which has an energy content comparable to gasoline. Boron is also totally safe to transport and store, and there is a quite plentiful supply of it on earth. There are of course challenges, as the boron needs to be burned in pure oxygen, which would need to be either carried along or be separated on the way from air. A boron-hybrid could also be used as emergency power, or used to help with electricity demand spikes. The same could be done at boron recycling plants.[29] The first tests with boron cars were done in Turkey in the summer of 2012.[30]

Aviation

> If the price per barrel rises to US$200 it would mean the collapse of the entire aviation industry. No aviation company can cope with that price.
>
> Thomas Enders, Ex-CEO of Airbus (2007)[31]

Flying is fast and cheap. It is cheap mainly because jet fuel is not taxed much, and because the industry gets rather sizeable subsidies and support from the rest of society. In practice this means that the costs of aviation have been externalized for others to pay. Flying is a bit like driving on steroids: one can go almost anywhere, but much faster and without congestion (except in airports). Many western citizens are surprisingly dependent on aviation, or at least they feel that way. Commercial flying has been transformed from an elite luxury to a routine of the masses in just a couple of decades. A lot of freight is also transported by airplane, but the focus is on freight value instead of weight or volume.

Flying is not much worse in its energy efficiency than driving. The main problem is due to the longer distances travelled. The amount of fuel that one long distance flight consumes per family would drive the family car for a year. A Boeing 747 burns about 1 gallon (4 liters) of jet fuel per second. During the flight, the consumption is over 10 liters per kilometer.

But as the plane can hold over 500 passengers, a full plane ends up burning around the same amount of fuel that a full passenger car does per person. It also moves 10 times faster.[32] Improvements in fuel efficiency for new plane designs are sorely needed. The average U.S. citizen gained 4.5 kg of weight between 1998 and 2008. This resulted in increased annual fuel use of around 1,300 million liters (350 million gallons) for the aviation companies.[33] To put energy conservation methods in context, the energy saved over a lifetime by switching all one's lights to energy-saving bulbs, can be easily burned on just one flight. Small actions will have small effects.

Aviation, in addition to driving, is one of the first industries facing problems in a world of more expensive oil. Often as much as a quarter of airlines' expenses come from aviation fuel. The fuel costs of flying will increase at the same time as increases in food and energy prices. Economic problems will also leave less disposable income for long-distance vacations. Many aviation companies have been in severe financial distress for years, with the new cheap flight companies making their situation even worse. In the future, it seems likely that the business will be consolidated into the hands of fewer companies, which will mean less competition. Less competition makes it possible for the prices to rise and expenses to decrease (flight crews will be more abundant which will show in their salaries). This will enable the businesses to operate for a while, until a new round of elimination is forced on the companies. Aviation will not disappear, but the current mode of mass flying with added services will face worsening problems.

Trafi, the Finnish transport safety agency, published an interesting report in 2012: "Finnish aviation up to 2025 – four scenarios," which offers four scenarios with interesting background assumptions for any western nation:[34]

- from a globalized world to a world of blocs
- Europe on its own course
- economic prosperity with reforms
- from oil crisis into economic recession.

The baseline of the report is that the age of cheap oil is over, and that the alternative futures will see 2 to 3 times higher oil prices than we had in 2012 (around US$110 per barrel). To quote: "It would seem a rather unlikely scenario that oil prices would be at the current levels in 2025." The proportion of fuel expenses at aviation companies has been increasing, from 13 percent in the early 2000s to as high as 30 percent by 2012.

The most extreme scenario, which is also the most interesting from this book's premise, is the one "From oil crisis into economic recession." According to this scenario, "High oil prices will have a negative effect on the economy. The price of oil will be three times higher in 2025 compared to the 2012 level," and the situation will unfold roughly as follows (slightly summarized):

The European economy will cease growing, and the new reality after failed reforms will be that of prolonged recession. The growth in Asia will also slow down, with the aid of the recessions in Europe and the United States. Exports from Asia will decrease significantly. Russia will become more unstable and also cease growing. Business communications will be done mainly virtually, as the price of traveling increases along with oil prices. Tourism will find a new trend in local and domestic traveling due to increases in the costs of flying.

In the scenario, long-distance flying will decrease, and the prolonged economic recession brought on by expensive oil will see many aviation companies going bankrupt. The scenario is framed strictly from the viewpoint of aviation. It does not even try to take into account the results of collapsing tourism on the economies and countries that depend on it, even though recognizing the likely decreases in traveling caused by recession. The world of "peak" is not very encouraging for the aviation business.

Shipping

Ocean freight represented over 80 percent of global trade, measured by volume, in 2008. Over 35 percent of ocean freight capacity is oil tankers.[35] Around one half of global oil shipments are transported via oil tankers, which means around 44 mbpd.[36]

Ocean transportation has, due to automation and standard containers, developed to be a rather cheap and efficient method of transporting goods. Fuel is of course burned, but compared with the amount of goods transported, the amount is rather small. In addition, even this amount can be cut significantly by reducing cruising speed slightly. Even wind power has been making a kind of comeback in ocean freight. There are many kinds of sails being worked on which would help cut fuel use in the open ocean.

More than 95 percent of international trade in ton-kilometers is transported as ocean freight, but it only accounts for less than half the value. Of total trade, both domestic and international, two-thirds is done via shipping in ton-kilometers.[37] The weakness of ocean freight is the relative slowness. It is also dependent on safe sea lanes. These have recently been made possible by the overwhelming naval power of the United States, even if the Horn of Africa does have a wide variety of international naval ships keeping the area more safe.

A decrease in cruising speed is one option to cut fuel use. The most effective, if not the most politically correct, way to cut fuel use would be to make small nuclear reactors more common in ocean freight. There are at least a few factors that support this:

- The technology has been proven in naval military ships and ice-breakers with a history of more than half a century, so this is nothing new. There have not been many accidents.

- With small reactors of some tens of megawatts of power, any accident would have quite limited consequences. In addition, oil could still be used for navigation near coastlines and harbors if needed.
- There would not be any small particle pollution.
- Cruising speeds would not have to be reduced due to fuel costs, making ocean freight more competitive with other fossil-fuel powered transportation.
- Fuel costs would drop significantly.
- Greenhouse gas emissions from shipping would be reduced to almost nothing.

According to one consultant in the ocean shipping business, whenever he mentions nuclear, the reception is often that of disbelief and wonder. Talking about biofuels seems to be more politically correct and popular, even though their negative side effects and problems with scaling are well known.

Railways

Electric rail is one of the most energy-efficient methods of fast transportation. This is due to the small air resistance, small friction and the great efficiency of the electric motor. A full electric train is comparable to cycling in energy use per passenger, even though the speed is nearly 10 times faster. An electric car packed full (four passengers), a full subway train, or a full tram are also about as efficient. Driving alone can use 50 times more energy than an electric train.

Building railways is both expensive and slow, so they demand a high rate of utilization to be sensible. A high utilization rate on the other hand requires that the people driving cars on the highway beside the railway can be attracted to change from the car to the train. It is easy to blame railway investments for being costly and ineffective. In the countryside, there is little sense in investing in light rail or local trains that will end up running mostly empty most of the time. It makes as little sense to drive around in personal cars in cities.

A major obstacle in developing European railways is the difference in standards on electrification and the width of rails between countries and areas. In the EU there are nineteen different communication standards in use that are not mutually compatible. There have been efforts to homogenize the system since the start of the century.[38]

Overview

Transportation and logistics are essential for modern society, so we have to manage them somehow. On the other hand, transportation is the logical place to cut our oil use and dependency, as more than half of our oil is used in it. These cuts seem inevitable for oil-importing countries, whether they

Prices of transporting goods per ton per 1,000 km

- $15 rail
- $180 road
- $551 air
- $10 pipeline.

Source: Gilbers, R., Perl, A. (2010), *Transport Revolutions*, New Society Publishers.

are done because of trade balance, an oil crisis or climate change. When oil becomes more expensive, road transportation loses some of its competitive advantage over other methods of transportation. It will still claim a piece of the pie, however, as it is not effective to build numerous alternatives to every destination. Fuel efficiency could also be improved for example with hybrid technologies, which would be nice to have in heavier vehicles. The easy fruits ripe for picking are increases in the utilization rate of the current fleet through different kinds of ride-sharing. These could be endorsed more by the state as well as by employers. Modern smartphones offer great possibilities through combining GPS location data and internet services. Regulation should also be loosened, since now even piloting these kinds of innovative services, even at the local level, soon comes up against a brick wall of regulations, laws, officials and bureaucracy.

Food production

> The most important factor for our future survival is food. Peak oil will force change in our systems of food production.
>
> Kjell Aleklett, *Peeking at Peak Oil* (2012)

The human population has exploded in just a few decades. The relative rate of growth has, however, dropped from the 1970s level of 2 percent to around 1.2 percent. With current trends, the population in 2050 will be between 9 and 10 billion.[39] A recent report by the UN revised these trends slightly upwards – it estimated that in 2100 there would be 10.9 billion instead of the previous estimate of 10.1 billion.[40] According to Professor Aleklett and many others, we can either feed the population, or feed our cars with biofuels, but not both.

Throughout history our population has remained rather stagnant at well under 1 billion, and the population is now growing by nearly 1 billion every decade. Fossil fuels are behind most of the reasons and factors that have made this growth possible.

While food production has increased massively, the share of people working in food production has decreased to just a few percent of the population. Only around 2 percent of people in developed countries work in food production. The situation was dramatically different just a few decades ago. Our society is based on specialization. This in turn is only possible because a small portion of the population grows the food for everyone else, so they can do something else. It is this surplus of food production that is also at the heart of economic growth.[41]

A modern farm can use more energy in production than is contained in the harvest, although with farms growing crops, this is an odd exception. When we take into account a varied diet with meat and dairy products and not just crops, and include transportation, packing, cooling, preparation and waste, we actually spend between 5 and 10 calories for each calorie eaten.[42] These calories come largely from fossil fuels. With a growing population and developing countries adding more meat to their diet, this equation will become more and more unsustainable. Producing meat can take much more energy than producing crops, since a lot of beef cattle are fed with crops instead of pasturing.

With growing costs in transportation, imported food will increase in price compared with local production. If we cut our use of fertilizers and pesticides, we will get smaller harvests which will also increase food prices. Meanwhile, biofuel production is competing with food crops for the same acreage.[43] We have quadrupled the human population in just a few decades. How were we fed?

Green revolution

During the twentieth century, the crop harvest per acreage grew many times over. This *Green Revolution* was accomplished through several key factors:

- plant breeding
- adding irrigation
- agricultural mechanization (e.g. tractors, harvesters)
- fertilizers and pesticides.

Plant breeding as such does not increase the use of oil, and can even decrease it. It has been hugely important in increasing harvests. One of the most important pioneers, Norman Borlaug, received a Nobel prize and many high national recognitions and awards.[44] But irrigation is done mainly with diesel-powered pumps, mechanization uses the internal combustion engine, nitrogen fertilizers are made from natural gas, and pesticides are made from oil. It is hardly an exaggeration to say that we eat fossil fuels that have been transformed to edible food with the help of the soil. Mechanization gives farmers super powers. A farmer in the U.S. has the power of 8,000 people through various machines. This still pales in comparison with the power of

28,000 harnessed by oil drillers.[45] In addition to the actual growing, modern food uses a vast amount of energy before it is eaten. Packaging is often made of plastic which helps reduce spoiling. The food is processed, prepared, and pre-cooked, after which it is cooled or frozen. Food is also transported back and forth thousands of kilometers. One of the most glaring energy uses is when the mother or father of the family takes the 2-ton SUV from the garage and drives to the nearest hypermarket to collect 0.02 tons of food. This can easily burn a couple of deciliters of gasoline per kilogram of food and drink. The food is then cooked, often with gas or electricity. After the family dinner, a large portion of the food that has been through this whole process is thrown in the bin while it is still edible.

The growth in harvests has been slowing down in recent years. More and more fertilizers and pesticides are being added, with smaller and smaller corresponding increases in harvests. Twice the amount of nitrogen fertilizers are used in China compared with the United States. In the U.S. farmers often use soy beans in crop rotation, as it adds nitrogen to the soil, reducing the need for added fertilizers.

Grain prices have doubled since the mid-2000s. This development has been enforced, as well as by higher oil prices, by a few other worrying trends:

1 Rising prices of oil increase the appeal of ethanol production from grain.
2 In addition to increasing population (about 80 million more each year) there are around 3 billion people aiming to add more meat to their largely vegetable diets.
3 The levels of groundwater tables are getting lower in big grain producing countries. In China alone, it is estimated that 130 million people are fed by over-pumping the water aquifers.
4 The harvests per acreage in many developed countries show signs of leveling off, so the past trend of ever-growing harvests is coming to an end. Many plants are facing the limits of photosynthesis and biology, which cannot be stretched much further by adding fertilizers or breeding.
5 Due to climate change, harvests will get smaller and damage due to insects and plant diseases will grow. As a rule of thumb, each 1 degree above the optimal temperature will decrease the harvest by 10 percent, but this estimate may be too optimistic.[46]
6 An area the size of Ukraine of arable land is being eroded each year due to drought, deforestation, and climate change.[47] [48]

A report[49] published by the UN in late 2013 came to conclusions about some of the fundamental changes we need to make in the way we grow and handle food through the whole production chain (the report only represents the views of the writers, not the UN). Greenhouse gas emissions from agriculture and food production have been growing for the last few years, even though they need to be rapidly declining. Agriculture will suffer from progressing climate change. It acts as a multiplier to other problems, and it will also bring a host of its own risks to bear. The report sees several changes needed in the way we produce food in addition to mitigating climate change:

- The incomes for farmers need to be better and more secure.
- We need to move away from monocultures that damage the soil to a more varied and sustainable food production on farms.
- The resilience of food production needs to be improved to better handle the challenges brought about by energy scarcity and climate change.

Agriculture needs to be seen more holistically than currently. For example, the potential for storing carbon in the soil needs to be further developed, as it will help mitigate climate change and enrich the soil. Waste in the food chain needs to be cut radically, and the use of organic and inorganic fertilizers needs to be optimized.

Problems in food production

As food prices rose by 50 percent between 2007 and 2008, many producing nations banned exports, so that their own citizens could afford sufficient food. This in turn led to panic in food-importing countries, as their trust in the global food markets eroded. This resulted in many bilateral negotiations to ensure food deliveries with long contracts. This in turn eroded the trust in global food markets even further. In addition, some rich countries have started buying and renting arable land mostly in Africa to secure their food production. This land-grabbing had amounted to more than 56 million hectares by 2011, which is more than the combined amount of land used to grow corn and wheat in the United States.[50]

Record amounts of fertile soil are lost yearly due to erosion, much faster than new soil is formed. Accumulating salt is another problem in some areas. That soil, which is used in industrialized agriculture, often acts as just a dead sponge where seeds are sown along with added water, fertilizers, and pesticides. In many places, microbial life that would keep the soil fertile and productive has collapsed to almost non-existence due to the continued use of chemicals. Regenerating the soil and its microbial activity, if it is started at some point, will take several years. The actions we can take now are always dependent on the actions we have taken before, and because of the slowness of the changes needed, this also applies to the near future.

The ever-increasing need for fertilizer use is the result of our efforts to replace the natural nutrition cycles with our system that runs on fossil fuels and sweetwater, namely flushing it down the toilet. With this system, the nutrients often end up in places where they would not be needed, such as lakes, rivers, and the seas. At the same time we use fossil fuels to mine and manufacture synthetic fertilizers to replace those lost. A person living in a large modern city is often quite removed from the reality that involves excrement. We don't often think about the fact, that, though we flush the toilet, the product does not disappear just because we lose sight of it.

The need for energy inputs can be reduced in several ways. Crop rotation and no-till farming are two examples. In no-till farming the soil is not disturbed by tilling. This makes it more fertile, better able to hold water, and resistant to erosion. It also uses less fuel. In crop rotation the fertility of the soil is maintained by changing the planted crop from time to time, and planting nitrogen-fixing crops such as soy beans. Crop rotation also helps with pest control, as the pests are less likely to find the fields if their favorite food keeps changing to another place. Drip irrigation, which is done directly to the soil, can be used to reduce the energy use of irrigation. There are plenty of better methods, but spreading them to wider, everyday use will take years or decades.

The ongoing, massive sixth wave of species extinction is a logical result of human activity. Monocropping decreases the amount of species that can live in one area. Permaculture planning aims to address this challenge for its part. It is a philosophy that aims to maximize biodiversity, along with harvests, from various plants by actively designing the areas to be smart and able to maximally use different mechanisms offered by nature. It has mostly been used in small-scale gardening, so its feasibility for large-scale, mechanized food production remains unproven.

The financial and banking sector also plays a role in agriculture. In large-scale agriculture, the yields are large, but so are the required financial up-front investments. These are normally funded by debt. If the financial system faces a debt crisis, it can be hard for the farmer to borrow money to buy seeds and other necessary inputs. Large-scale agriculture is therefore susceptible to the whims and crises of the financial system, as is most of our economic activity. It is a slightly scary thought that the global production of food is so dependent on something that is so far removed from reality as the global financing markets.

European food production in a nutshell

Around 1.7 million square kilometers of land is used in agriculture in the European Union. This is about 40 percent of land, and around 3,400 square meters per citizen. Three percent of this, or 5.3 million hectares, is farmed organically. Around 60 percent of the area is in annual crop production, a little over a third is in pasture, and around 6 percent is in perennial use (this is concentrated in the Mediterranean area). The main produce in 2010 was wheat (140 million tons), corn (60 million tons) and barley (55 million tons).[51] Almost half of the European livestock was cattle, 27 percent is pigs, and 14 percent poultry. There were around 134 million units of livestock.

Around 25 million people worked in agriculture. They made up a total of 10 million human labor years (the EU has around 500 million inhabitants). A large proportion of the people working in agriculture do so only part-time. There were around 12 million farms, of which half were small farms of 2 hectares or less. Three percent of the farms were over 100 hectares, which represented more than half of the total agricultural land in production. The monetary value of agriculture in 2010 was around 280 billion euros[52] (360 billion US$), which is a little over 1.5 euros (2 US$) per citizen per day. In 2005 the value of food exports was 154 billion euros (200 billion US$), while the value of imports was 179 billion euros (230 billion US$). Net imports were therefore around 25 billion euros, or 14 eurocents per citizen per day.[53] Cheap transportation has enabled the transportation of similar foods between countries, even though it makes little sense from the perspective of energy efficiency. A typical consumer can also take a long look in the mirror, since there would not be imported cookies by the millions, if we bought domestic cookies. People have grown accustomed to imported delicacies over the years, and the demand is likely to continue into the future.

Agriculture's share of the total GDP is rather small. What people often seem to forget is that without agriculture there would be no other economic activities either. The same is largely true of oil and other critical energy sources and raw materials that are hard to substitute. Without them, there is no economy. No manufacturing, no transportation, and no services.

Fishing is one of the most energy-intensive ways to produce food. The global fishing fleet burned 1.2 percent of global oil in 2000. For each kilogram of fish and shellfish, 0.67 liters of fuel was burned. During the last decade, fuels costs have soared by 80 percent and the catch has dropped (largely due to overfishing) by 17 percent between 1995 and 2002. It is no wonder that the European fishing sector is in financial distress.

Spain (16 percent), Denmark (14 percent) and the UK (13 percent) together account for 43 percent of the European catch. Since 1995, fishing has decreased by a third. In 2010 the catch was slightly above 6 million tons. About 1.3 million tons of this was farmed fish. The European fishing fleet has around 84,000 ships. The fleet weighs around 1.75 million tons with a total engine power of 6.5 million kilowatts. In 2008 it burned 3.7 billion liters of fuel.[54]

U.S. food production in a nutshell

The U.S. has more than double the area of agriculture compared with the EU, totaling 3.7 million square kilometers, or 11,600 square meters per citizen. The United States is a significant exporter of agricultural products, and it has been called the bread basket of the world. Around one-fifth, both in value and in volume, of U.S. food production is exported.[55] The main crops are corn, soy beans, wheat and forages such as alfalfa. In 2007 there were around 96 million cattle, 67 million pigs, 6 million sheep, 1,600 million broilers and other meat chickens, and 350 million egg-laying hens.[56] The value of meat production in 2010 was 78 billion US$ (60 billion euros) and the value of farm products in 2009 was 299 billion dollars (232 billion euros).[57]

There are several threats to U.S. food production and exports. The most direct is climate change that can result in worsening droughts and an increase in crop-destroying storms in agricultural areas. Another threat is the biofuels policy, which threatens to take an increasing part of the corn harvest to be made into ethanol. This would increase global food prices.

What about organic farming?

Moving to organic production methods reduces the need for external energy inputs, but it also often decreases the harvest per area. Moving to organic production is not without its problems, and the harvests can be quite abysmal in the first few years of production. This is due to lack of knowledge about the new production methods and due to the soil usually being in rather bad shape after conventional farming. For a single farmer, a move to organic production can be impossible due to economic reasons. It is probably not possible to face several years of sub-average harvests economically. In addition, there are a few years of quarantine to be applied to production areas before they can be labelled as organic. Because of all these factors, it is not possible to quickly move to organic production on a large scale without severe problems with food availability.

Organic certificates, and strict regulations that come attached to them, have both positive and negative effects. Standards exist for organic production, in principle, to ensure that certain norms are followed and the consumer is getting what they are paying for. It is however possible to get round the limits on chemicals and tight regulation by forging the information on what substances have been used in the production. Regarding the European market, it is sufficient for a product to get inside the borders, and it is unlikely that customs supervision is equally strict in all EU countries.

Strict regulation causes people to make a steep division between "good" and "bad" production in their minds. In reality there is a vast range of production between conventional and organic agriculture that can use much less energy and have fewer negative effects than those of conventional, or even organic, agriculture. This information fails to reach the consumer, since the organic brand has taken over this segment. There is also limited space at stores to add information in addition to the Fair Trade and organic symbols.

Genetic engineering, for example, has been used to make crops that require fewer pesticides and fertilizers but which still have high levels of harvests. They can also handle the strain caused by climate change better. In the best case, a combination of organic styles and genetically modified organisms (GMOs) would bring good harvests with minimal inputs of chemicals and energy.

Genetically modified organisms

> The scientific research conducted so far has not detected any significant hazards directly connected with the use of genetically engineered crops.
>
> Conclusion of a review on the last 10 years of scientific GE studies[58]

Europeans have so far been able to be squeamish about their food, where it is from, and how it was grown. This has also resulted in opposition to genetic engineering (GE) and genetically modified organisms. Nevertheless, genetic engineering could be a partial solution, as it can be used to reduce the amount of external energy needed to grow crops. It could also increase food production possibilities on less fertile marginal lands and climates, and help us cope with the negative effects of climate change. It can also be used to improve nourishment for the poor who lack sufficient vitamins in their current diet, as with "Golden Rice."[59] Genetic engineering has its own drawbacks, however, which need to be balanced with the potential benefits. Many of the problems that have been linked with genetic engineering actually have more to do with the multinational corporations and their practices, Monsanto being the most prolific example. People are being anti-GMO, but they are actually protesting against corporate mentality, profit maximizing, greed and the weaknesses in our current finance-based capitalism. These have little to do with genetic engineering itself. A lot of the opposition is also due to the hysteria and rumors that spread through the internet and social media. Most of the time they have very little in the way of actual scientific evidence behind them.

The scientific consensus on the assumed dangers and harm of genetic engineering is quite clear: there are no significant hazards connected with the use of GMOs.[60] [61] [62] [63] [64] [65] One of the more worrying manifestations of anti-GMO activism has been the sabotage of further studies, both privately and publicly funded. The purpose of these studies is to get more information on the possible benefits and dangers of genetic engineering, so there would seem to be little to be gained, but a lot to lose, by sabotaging them.

The critical public opinion has been largely responsible for driving GE research from public universities and research laboratories into the laboratories of big agri-businesses. This causes the aims of GE research to be more focused on profit maximization than they would otherwise be. Serving society as a whole becomes less important.

Central questions

It is common to have several points of view when discussing a nation's own food production, and so it is easy to get bogged down in a jungle of misunderstandings. The discussion could be framed by inspecting food production, with its energy flows and effects as a whole. After that, the discussion could be widened to include the level of national food security and independence from imports and economics. What options are available and what are their time frames? Below are few central questions to get the discussion started:

1. How much trust should we put in the global food markets and their smooth operation in the future?

   As the world could well be heading for more crisis instead of less, it might be difficult to ensure an ample supply of grain and such from the global food markets. If there are severe losses in crops, the nations will first ensure the food supply for their own citizens. If we want to buy food from the spot market or with short contracts from wherever it is the cheapest, it might mean uncertainty in availability and prices in the future. How do we control this risk?
2. Who pays for the possibly bad economics of domestic food production?

   Certain food stuffs, depending on the country, are always cheaper to grow somewhere else. What are these, their volumes, and who pays the bill if they are to be produced domestically instead of imported, and with what mechanisms? The EU currently has different kinds of agricultural subsidies, but these have their own political problems. At the store, the price is often the deciding factor for the buyer, so the domestic food cannot be significantly more expensive.
3. How do we reduce the imported inputs of domestic food production, or do we even want to reduce them?

   Domestic food is often produced with imported fossil fuels and other inputs. Is it sensible to try a shift to a more organic form of agriculture? Who will pay for this? And what value does the smaller dependence on imported energy inputs have in the future?
4. Most of our current agricultural methods are very dependent on mechanization, 24/7 repair services, and a constant supply of energy both in electricity and liquid fuels.

   What risks does this bear in the future, as fuel becomes scarcer, more expensive and the risk of supply disruptions grows? If domestic food production is seen as essential, how do we ensure that it gets the energy it needs to function, preferably without going bankrupt?
5. How permanent is the current trend where only a few percent of the population works in agriculture?

   In just a few decades, with the help of the Green Revolution, we have totally changed the economic structure of society. If the role of fossil fuels is larger than commonly thought, and those fuels become scarcer, this trend might turn. How fast can the changes needed occur? How should it be managed in a controlled manner? What does this managing mean for the economy?
6. We also face the choice of producing biofuels or human food from our crops.

We do not have the capacity for both. What is the realistic time frame and volume for when second and third generation biofuels become a reality? Who will make the decision between biofuels and food, on what grounds, and how will the decision be implemented without significant negative side effects?

The structure of society and cities

The increase in cars and driving has led to society being structured on the premise of the car. This can be observed as superstores and service stations keep popping up at the intersections of highways. People of course often like the wide selections and cheaper prices of superstores. It is hard for the smaller local stores to compete. The size of the shopping basket has also been growing. This is good from the transportation side of the matter, as there will be less driving to the store. It is bad because more food is likely ending up in the waste bin due to spoiling.

As the infrastructure of society has been spreading through the fields and forests, the availability of the car has become more and more a necessity. A large part of our oil use has been made inelastic through the structure of our societies. This means that we simply have to buy gasoline, no matter what the price. In many places there has been no sign of an upper limit on prices that would instigate people to stop driving. Driving, and using gasoline as the fuel for it, is seen as essential. The money that is spent on more expensive gasoline is taken from somewhere else. Usually it means that a smaller part of our income stays in domestic circulation.

Zoning and building permissions

We live in a society that offers us many services. When one builds a house, it means that resources are invested in infrastructure that will be in use for decades, and usually not just for the one who built the house. If the rest of society is obligated to deliver services and infrastructure to almost every home, it seems reasonable that society also has a say in where people build their houses. Electricity, roads, water and wastewater, postal services, schools, internet connections, and numerous other services and benefits are delivered to almost everywhere, at least for some, usually at a subsidized price. Even if the one building the house promises to pay for the infrastructure himself, he can only speak for himself. The next occupant in the house has not made any such promises, and it would be wasteful to build houses that would remain empty. From the point of view of society and its resource use, it matters a lot where and how people live.

A further problem is the current lack of knowledge about peak oil (and other trends) among the public. How much desire would we have to build that big manor in the middle of nowhere, far from services and the rest of the society if we were aware of the following:

- Fuel prices will keep on climbing. Oil prices are set to double in 10 years according to the IMF[66] and OECD.[67]
- Our economic situation will be far worse than has been estimated.
- The government will not have the resources to subsidize the expense of driving, and cities and communities will have to cut and centralize their services.
- The availability of liquid fuels could be severely limited much faster than many of us could imagine.

With the knowledge of these possible future scenarios, the desire for car-based municipal and regional development could be greatly reduced.

It is of course a different matter if someone moves to the rural area to gain their livelihood from agriculture, forestry or other such rural activities. The most problematic people are those who live in the cities but go to sleep in the so-called "bedroom communities" in the countryside. Vast amounts of energy are used just for the daily commute. All of this has been possible only because energy has been cheap.

If we aim to produce food in the future, we cannot empty the countryside of people. Nobody will want to live in the middle of nowhere, with no people or services within tens of kilometers. The location of rural towns and villages, nearby transportation routes, and other infrastructure will determine how viable they will be. On the other hand, cities are utterly dependent on the surrounding (and even global) countryside for food, energy and other resources. So far the possibilities for improved life in the cities have been a big enough reason for many to leave poverty and subsistence farming behind. But the economic crash in some European countries has already seen a reversal in this trend, as the cities have failed to offer any meaningful livelihood.

Zoning is probably one of the slowest ways to increase society's energy efficiency. It is usually neither possible nor cost-effective to abandon current infrastructure. Designing and building new structures will take decades and it will be so expensive that there is little sense in doing it faster than needed. Even so, the new infrastructures are always built with the needs of the old infrastructure in mind. The availability of cheap fossil fuels has distorted the premise for design and zoning for decades. With peak oil and climate change at hand, we will be forced to pay much greater attention to energy efficiency in the future.

One sensible and practical solution would be to set energy efficiency as a top priority in all new zoning policies. The current infrastructure should be developed to be more energy-efficient and resilient. Ceasing the construction of new motor highways, zoning new development near current railways, using inland waterways more, and zoning commercial, industrial and residential areas to be nearer (from the point of view of energy usage) to each other could be some ways to reach these goals. Any progress will demand consistent policies and rooting out projects that are worsening energy efficiency significantly.

It could be argued that if people live less densely, the big yards and additional space could be used to produce food with gardening. This is a reasonable argument, but it needs to be put into context. Not many people are growing their own food even though they have the chance. Instead, many resources are used for gardening that produces nothing (think lawnmowers) except maybe beauty (which is worthy in itself but may be less so in the context of this book). So counting on the suburb as a major source of food anytime soon would be irresponsible. But it will be nearly impossible, and wasteful, to abandon the current suburbs since they have already been built. We must simply take the advantages from them that we can. With increasing food prices, it will become feasible for many to save significant parts of their income by growing and gathering some of their food by themselves. For now, the highly trained specialists are probably better off using their working hours to apply their expertise in other productive activities.

Oil in energy production

Oil has had a share of around 37 percent of Europe's primary energy use, while the global level has been slightly lower at around a third. Just by focusing on this number, the role of oil might seem at least somehow controllable. There is a wide variation between European nations in their oil use. Malta makes 100 percent of its energy with imported oil, while Estonia has oil at a share of 20 percent of primary energy. The oil use in Europe has been quite stable since the early 1990s. The share of transportation has been growing steadily, except for the recent financial crisis.

Three-quarters of European energy is from fossil fuels, while the global average is just over 80 percent. A significant portion of our primary fuels is produced and transported with oil. Europe produces less than half of the energy it uses, and almost 85 percent of our fossil fuels are imported.[68]

Oil's important role as the supreme source of secondary energy in the form of liquid fuels is often forgotten. While it is quite easy to produce

Table 4.6 Primary energy sources in EU

EU primary energy	*1990 (%)*	*2009 (%)*
Oil	38	37
Coal	27	16
Natural gas	18	24
Nuclear	12	14
Renewables	4	9
Total (oil equivalent)	1,665 Mtoe	1,703 Mtoe

electricity in many ways, liquid fuels are much harder – with the exception of refining them from crude oil. Liquid fuels have a unique role in modern society and everything that happens in it. The production of other primary energy sources is often directly or indirectly dependent on oil.

Fuels such as coal, peat or wood are the most dependent on oil, as their mining, harvesting and transporting runs on oil and the volumes are large. Bioenergy is probably the most dependent on oil use. Harvesting and transporting energy wood takes 3–6 percent of the heat energy contained in the wood. This energy is used in liquid fuels, which are much more refined and valuable than wood chips.[69] If we first synthesize liquid fuels from biomass (at a certain efficiency and EROEI) which we then use to harvest and transport more biomass, the weaknesses of the total cycle become apparent. Bioenergy is often subsidized through cheap oil, which is often forgotten when it is presented as an alternative to fossil fuels.

With imported fuels, most of the oil is burned in the exporting country. The price of oil will then indirectly affect the prices of fuels. Nuclear, wind, hydro and solar all use less oil. The main uses are in materials such as concrete and steel, building the facilities, and the necessary infrastructure.

It remains difficult to cut oil use in energy production. And if we aim to increase our use of liquid fuels derived from biomass, the oil use will more likely grow.

Oil in the economic system

Oil is essential for our modern society. Without oil, people are not able to satisfy their basic material needs. Economic or political activity beyond one's own village would become nearly impossible. Specialization is one of the key pillars of modern society. It is only possible on a large scale through money (which acts as a means of payment) and oil (which allows goods to move around, searching for the cheapest unit of labor).

The links between oil and the economy are clearest for many from the oil and energy crises of the 1970s. Back then, the western economies, built on their careless trust in cheap and available petroleum, fell to their knees quite quickly. A large-scale program of weaning businesses off oil soon followed the state efforts to ration fuel use. Oil soon ceased to be a major fuel for electricity generation.

Around the same time another, possibly even more economically significant, event happened. Saudi Arabia, and soon the remaining oil of the big oil producers as well, moved from using the gold standard to using the U.S. dollar as the only currency in oil trade. A few years before, in 1971, the dollar had been separated from the gold standard. The resulting petrodollar system enabled for its part the rapid growth of the global economy, as it offered a common currency for international trade and a growing reserve of dollars for the global banks to base their lending on.

Oil in international trade in numbers

- The share of oil in global ocean freight capacity is 35 percent.[70]
- About half of the world's oil use (44 mbpd) is shipped as ocean freight.[71]
- The total value of trade in crude oil in 2000 was around 450 billion dollars. In 2008 it was over 1,900 billion dollars.[72]
- With oil prices at 70 US$/barrel, around 1,000 billion dollars' worth of wealth is transferred to oil-exporting countries annually.[73] With 2012 prices, the yearly amount is between 1,500 and 2,000 billion dollars.

More oil use statistics can be found at Appendix 1 "Oil in numbers."

As the demand for oil remains strong and the amount of oil ending up in the global markets falls, the price will continue to rise. Since oil is an important factor in production almost everywhere, there will be inflationary pressures on the prices of most goods. During the twenty-first century, oil prices have risen 16 percent annually on average.[74]

According to neoclassical economic models, the elasticity of energy production should be the same as the value of the energy from the total GDP. For example in the United States oil has around a 4 percent share of GDP. This would suggest that a 10 percent drop in oil consumption would decrease GDP by 0.4 percent. The consequences of real-life oil crises have been 10 times larger, at around 4 percent.[75] With a peak in oil supply, the situation will get tighter. The demand for oil is inelastic, which means that it is hard to reduce oil consumption when prices rise. When the supply of oil also becomes inelastic due to the proximity of peak oil, meaning that production does not grow even if prices increase, we find ourselves in a new situation. In an inelastic market, the price fluctuations caused by disruptions can be 10 times higher than in elastic markets.[76]

The correlation between oil use and economic growth

When discussing the economy, we usually end up talking about GDP. This is natural. When discussing a complex system such as the economy, we have to simplify and choose terms and concepts that everybody knows and understands, at least on some superficial level. This results in a distortion of the discussion (and the studies, statistics, policies and other choices and actions), as the discussion starts to circle around the new gravity point, in this case the GDP. This happens whether the chosen concepts are actually useful in describing the problem, solution, or decision being made, or not. In the end, GDP is only a measure of the volumetric scale of publicly made monetary transactions happening in the economy. It does not measure the amount of activity in an economy, value creation, sustainability, well-being, justness of the system, or the actual wealth.

GDP, or the annual measured change in GDP, offers a neat and simple metric that at least appears to be scientific and accurate, that can be used to justify economic decisions. Upon closer inspection, this justification tends to make all decisions economic, even if they were actually something quite different. It is important for politicians to keep the GDP growing, since it is usually understood to add to the wealth of the nation, increasing tax revenue, profits, employment and viability. The growth in nominal incomes makes it possible to go deeper into debt, without seemingly going *too much* into debt.

The relationship between oil consumption and economic growth has been studied quite extensively, and a strong correlation has been found. It is of course possible that the correlation is explained by a third, unrelated factor, but this is generally seen as unlikely.[77] Different sources have estimated the direct effects of oil prices on the GDP of a typical oil-importing western country:

- The IMF and Goldman Sachs have similar estimates: as the oil price increases by 50 percent, the GDP in OECD nations is reduced by 1–1.5 percent.
- An article on the price elasticity of oil by Robert Hoffman says that a 10 percent increase in oil prices reduces GDP by 0.2 percent in the first year and 0.5 percent the second year. In the same article, a study by Hamilton is presented, according to which a price change of 10 percent between the years 1949 and 2005 resulted in a 0.7 percent drop in GDP.[78]
- According to an analysis on the Oilprice.com website, if the share of oil is 4 percent of GDP, a 25 percent increase in price would decrease GDP by 0.8 percent in one year.[79]
- Robert Hirsch, who was the author of the 2005 report to the Department of Energy on the effects of peak oil, has said that changes in oil consumption cause changes of comparable size in GDP.
- According to Jeff Rubin, a former chief economist at CIBC World Markets and author on peak oil, the ratio is 1:2. One percent more of oil consumption increases GDP by 2 percent.[80]
- According to Merrill Lynch, the energy sector represents around 9 percent of the global economy. According to Deutsche Bank, the global oil trade accounts for around 5 percent of global GDP.
- It is quite commonly thought that the global economy can withstand oil prices at around 120 US$ per barrel, after which it falls into a recession.

It seems likely that these estimates only account for the direct price effects of oil. They may not account for the destructed demand that those higher prices cause. This destroyed demand could have resulted in economic activity if that oil had been available. This effect, as it is hypothetical, is difficult to measure.

The historical effects of oil prices on the economy have been quite straightforward. When oil accounted for less than 2 percent of GDP, the U.S. economy was growing briskly (as in the early 1970s). When the share rose to nearly 4 percent of GDP, the economy started to slow down (as in 1973–1983).[81] Robert Ayres and Benjamin Warr have written more on the relationship of energy and the economy, for example in their book *The Economic Growth Engine: How Energy and Work Drive Material Prosperity* (2010).

The oil intensity of the economy

The concept of marginal utility helps to understand why the developing (non-OECD) countries are growing their share of global oil use. To simplify, by using one extra liter of gasoline in his aggregate, an Indian artisan can create much more additional value than the average European consumer, who will use that liter of gasoline by driving more during his summer vacation. The lower starting level of developed countries indicates that they can cope with higher prices, at least when it comes to using oil in productive work. This in turn indicates that the level of oil use in OECD countries has a long way to fall.

Decreasing oil use does not mean that we will simply return to things as they were before. The "before" does not exist anymore. A large part of the old structures have been destroyed as society, technology and economy has progressed forward. For example the superstores, made possible by the passenger car, have largely killed the smaller corner stores in many places. The same thing happened when steam ships replaced sailing ships in the oceans of the world. Society has been climbing ever higher on a ladder and breaking all the steps below. The way downward could therefore prove to be something other than a controlled and even descent.

The following are some examples of the old structures and technologies that have been replaced during the last few decades:

- Corner stores have been replaced with bigger markets and malls that are further from their customers.
- Temporal storage and storage at stores has been replaced by just-in-time deliveries.
- Cellars and other cool storage has been replaced by electrical fridges, freezers, and outsourcing storage capacity to the local supermarket.
- Social safety networks of communities have been replaced by social and health services offered by the government.
- Small local schools have been replaced by bigger schools that are farther apart.
- Horse-powered agriculture has been replaced by industrial-scale, diesel-powered mechanized agriculture.

- Local construction has been replaced by imported materials and elements made in a factory on the other side of the country.
- Home appliances and their repair networks have been replaced by integrated appliances that are expensive or impossible to repair.

There are also some clearly positive trends. Carrying letters and books has been replaced with ebooks and email. The growing trend of 3D printers could make the local production of spare parts possible once more, and so forth.

A quite common view on peak oil research is that the depletion of oil production will result in a jagged contraction of the economy. Structural changes that increase oil and energy efficiency make growth possible for a while, but then another crash is faced due to depleting energy and as competing economies make even more improvements in their efficiency. Long-term planning will become ever harder.

Direct effects on the economy

In addition to cost inflation, recession, wealth transfer to oil exporters and relative increases in developing nations' oil use, what other effects will there be in the global economy?

Since energy efficiency has not been very important due to cheap fossil fuels and their externalized costs, it has been largely neglected. As the prices of energy, transportation, and materials increase, energy and material efficiency will become much more central to all activities. The manufacturing of certain goods can move back closer to the consumer, and the whole manufacturing chain might once more be located under the same roof, or at least close by. Transporting materials and components back and forth will become more expensive and risky. The increasing uncertainty of deliveries makes local warehousing more competitive and the capital invested in bigger stocks of goods and components can offer an advantage where the just-in-time deliveries and lack of warehouses now offer one. This will mean that production chains will get shorter and that globalization will go into reverse.

Increased risks of deliveries are part of the changes brought by a world of peak oil and scarcity of resources. In this world international conflicts, as well as national conflict between societal or geographic groups, can become more common. This could mean that our just-in-time society will change into a not-in-time society. This has wide repercussions on how businesses operate. They have adopted almost borderless trade and binding international trade agreements as part of their normal operation, and they have optimized their storage capacity and other activities accordingly.

The U.S. dollar is a story in its own right. In addition to the fact that the United States has the monopoly on releasing more dollars to the market it is also the only official currency of oil trade and the reserve currency of the world. The U.S. government has seized on this, as it has consumed more and more at the expense of the rest of the world, and at the same time managed to pile an unprecedented amount of foreign debt. If the oil exporters were to move to other currencies, the consequences could be dire. The result could be a vicious circle where:

- The share of US$ in global trade (and oil trade) decreases, and other currencies increase their share.
- As a result, the need and demand for dollars in oil-exporting countries, foreign funds and currency reserves of central banks decreases, which causes the dollars to start returning to the United States.
- The reduced demand for the dollar reduces the demand for U.S. debt obligations, which will increase the interest rates on U.S. debts.
- This increases the need for the Federal Reserve to print more dollars, which it can use to buy debt obligations from the federal and state governments. More and more people find out that U.S. will service its debt through inflation, which erodes the trust in the dollar as a safe haven.
- As dollars return to the U.S. it causes price inflation.
- With higher inflation expectations, interest rates rise which makes servicing debt more expensive. This then puts more pressure on the Fed to print more money, and the money will start circulating faster in the economy, causing more inflation. And so on . . .

This would mean that the ability of the U.S. markets to stimulate production and investments globally by consuming more would be gone. Investments by foreign companies in the United States could become a problem and threaten the balance sheets and capital adequacy of many companies. The structures built assuming a long-term horizon cannot change overnight, so the result could be a structural change on a global scale. Behind the curtain, preparations are already underway. China has made huge credit agreements with various countries. It is also launching its own global currency transfer system (China International Payment System). China is also involved in creating the BRICS development bank together with other BRICS countries, which will transfer their dollars to infrastructure investments. On top of these, it is possible to build international trade and product chains that would be independent of the dollar.

Debt-driven consuming will have to change

Western consumers have been transferring their ability to consume tomorrow into their ability to consume today. This has been done by going deeper into debt, which has resulted in more economic growth in these countries than would have otherwise been possible. Increasing levels of debt, perhaps counter to normal intuition, has widely been considered as a positive thing and a sign of a healthy economy. This kind of consumer behavior will face severe pressure that can turn the whole situation on its head. People's willingness to take on more debt has been crumbling, even if the interest levels have stayed quite low. It is also possible that as the awareness of the oil production situation spreads, the resources used for consumption previously will be directed to investments that help reduce energy and oil use. Our debt-fueled consumption party and the businesses dependent on it will therefore suffer a double hit.

The whole consumer sector, including many companies, is between pincers: structural changes in the economy threaten to crash current streams of revenue. On the other hand, the cost inflation of critical resources and raw materials, and the austerity measures taken by governments will erode the purchasing power of the dollar or euro that can still be somehow collected from the market. There will be both inflationary and deflationary pressures on the economy and monetary system. Instead of reversing the effects of each other, together they make predicting and controlling the system all the more difficult. Contrary to gradual economic growth, the downturns tend to be fast and furious. Those companies that are more dependent on the mass consumerism need to be especially adaptable.

When in recession, companies compete for survival by adapting most efficiently to the changing environment. This is called creative destruction. This time, cheap oil is not coming back to restart the economy and growth, so this destruction may continue for much longer than we are used to. Creativity can still be found, but mainly in the explanations that economists, analysts and politicians offer in trying to shift the blame of the problem away from themselves. Different countries seem to be managing the transition to the resource scarcity with different degrees of seriousness. China for example has been ensuring its future sources of energy and raw materials consistently for years.[82]

Less is more

Up until now, the common thought in many countries has been that increasing population, sometimes through immigration, will improve their economic situation and make the demographics more manageable. In a world of increasing resource scarcity, this premise is turning on its head. If a country imports more than it exports, a growing population will not help the situation. This in itself does not change the relative use of resources by different sectors. The export industry will consume its share and bring export revenue, and sectors using imported resources will still consume their respective share of imports, and these have to be paid for. In fact, if we face various peaks in resource production, then more population means less to be shared among everyone. This holds true both in the local and global sense. With increasing resource efficiency this situation can be mitigated, but it does not change the basic arithmetic. Growth economics works poorly when the world is full.

As long as the new, young consumers base their consuming mainly on imported goods and resources such as oil, the effect they have on the trade balance can even be negative. New population in itself does not increase the availability of natural resources or the export revenue received from them. The structural changes ahead will likely result in lots of unemployed people in many places. It is doubtful that many of them can be employed in the export sector. This will force cuts in consumption to fix the trade balance that will be deeper the greater the population. In a world with limited availability of critical resources, less is often more, at least up to a point.

Oil markets and societal policies in oil scarcity

One definition of the market is that it is a mechanism for price discovery. The oil market consists of marketplaces where buyers and sellers trade in notes and securities representing oil and services, businesses or products based on or related to oil or oil usage, with prices that they think are appropriate based on their knowledge and assumptions about the future. The above sentence might be complex, but so is the reality it is trying to frame.

In spite of its weaknesses, the market is a good way to share resources between people and communities. It however needs to be remembered that the markets are based purely on prices. This means that the diversity of values, rights and responsibilities of the human race needs to be somehow measured in money. There are always groups that think that current market prices are not a just way to share the possibilities to consume. The values and rights of every human (let alone the rest of the environment) are never priced correctly. The size of this group is directly correlated with the inequality of the society. A popular survival strategy in many European crisis nations has been to leave taxes unpaid. This reflects a distrust of the government and their ability to use taxes for the common good.

When oil becomes scarcer, cuts in consumption need to be made. The market handles this by increasing the prices. This, according to market theories, causes people to switch from consuming oil to consuming something else. It is more likely that various special interest groups will see the situation as unjust, and lobby the government to direct the cuts in oil use to happen somewhere else.

This kind of interference by lobbyists and the government confuses the markets. It creates a threat that any political decision can be undone by another on a whim. It also creates a parallel system with the market, where the ability to consume is distributed according to political needs and wants. This leads to all able groups trying to maximize their share within the boundaries of the current regulations and rationing, which leads to further disturbances in the situation as a whole. The ensuing bureaucracy will need governing and new legislation, which will lead to a growing public sector and its share of consumption of resources.

In reality, the government takeover of the marketplace is well prepared for bureaucracy. The members of the IEA are required to have at least 90 days of net imports in emergency oil supplies. In addition, the IEA has various protocols ready to be implemented in case of an oil shock. These can be divided into two sections:

1. Increasing supply by
   a. releasing oil from the emergency supplies
   b. increasing oil production (short-term, spare production capacity).
2. Decreasing consumption by
   a. light measures such as information campaigns, eco-driving, and car-sharing
   b. medium measures such as lower speed limits on motorways
   c. heavy measures such as bans and limits on driving
   d. substitute fuels (mainly in electricity production).

Recently the IEA has twice enabled the release of emergency supplies – during Hurricane Katrina and the Arab Spring. As the market becomes more inelastic due to peak oil, the temptation to use emergency supplies for price control will grow. Temporary supplies are not suited for dealing with long-term trends however. On the contrary, they can be used to mask the underlying severity of the situation for a while, which will eventually make it worse.

When the state takes control of the marketplace, there will be a new deal of cards in society, with new winners and losers. In reality, most of the winners are those who suffer the least, so their "victory" is only relative.

Relative winners in oil scarcity

The first winner is the public sector and government. The majority of activities include governance and bureaucracy that in themselves do not produce anything. The government, with the bureaucrats and officials, are the ones who make the decisions on how the oil is divided, and they naturally see their own function and value as very significant and important. This means that the government and those close to it will get relatively more.

Other winners are the loudest special interest groups. It is typical for them to create a false image of importance around themselves and their activities. Politically linked groups and organizations can also be counted here, as they will search for loopholes that will ensure their special treatment in the new regulation and legislation.

Price inflation on the essential goods and energy will hurt the poorest the most, since they spend a bigger part of their income on these as is. Depending on the ruling political party, bread will be distributed more or less for the poor, and the bread will be acquired more or less fairly from somewhere else. The poor can hardly be called winners, but they do have a much shorter distance to fall. Their numbers will likely grow significantly.

The export sector is one of the winners, and with good reason, since their export revenues are used to pay for the imports everyone needs. This sector will therefore get the oil it needs. There will also be incentives for companies to get a domestic competitive advantage with the rationed oil instead of using it to increase exports.

The transportation sector is also essential for national economies and society. The amount of oil it uses drops automatically as the economy suffers and fewer goods are moved around. Sufficient oil for critical goods and deliveries has to be ensured so that society remains functional. On the other hand, nobody seems to know what and how much those critical goods and deliveries are.

The army and other national security institutions will face conflicting interests. On the one hand, shrinking budgets will be cutting resources that are used for training. On the other hand, these organizations will find it easier to justify their existence and needs in a seemingly more insecure world.

People living in rural areas, especially if they are involved in the resource-producing industries such as forestry and agriculture, could also be among the winners. Increasing self-sufficiency is easy to argue as the availability of food, energy and raw materials from the global markets gets more expensive and less secure. On the other hand, the longer distances to everywhere in the countryside will play an even bigger role with oil scarcity.

So if almost all oil users seem to be somehow privileged, is there a clear loser to be found? Much oil is used today with no regard for tomorrow and with little actual value created. The problem is that from the perspective of these users, a lot of value is created, and any rationing of this use will be seen as a violation of their basic human rights. Probable losers are the leisure and hobby industries. It might be worthwhile to enjoy the power-boats, sports cars, traveling, and karting while one still can. There might also be limits on work trips and commuting no matter how much money one makes. How does a negative reimbursement on the use of a private car sound?

The decisions and actions taken by the government always have some hidden and often negative side effects. If there is not enough information and the situation unfolds quickly and by surprise, these effects will be even worse. Nicole Foss, who analyses peak oil and economic crises, has stated something along the lines that ". . . in a crisis, the role of politicians is to make the situation worse as expensively as possible." Of course the reasons for any personal failures can always be found from somewhere else.

In the case of a severe supply shock, the first reaction goes to national emergency supply centers. With their supplies we can run our society more or less normally for several months. This would allow time for preparations to announce a national emergency or martial law, which will give state officials wide authority to direct the activities of citizens and businesses. This enables, for example, the rationing of liquid fuels and the limits on driving. As the situation continues, it also enables the confiscation of fuel-efficient vehicles from private citizens for the state to use in their critical deliveries.

The former life and business-as-usual will cease to be, and will soon be just memories as we will struggle to adapt to the new normal. Winners, if such can be found, might be the security sector and the army, which will move to a more central role in society.

Global security

> Where-ever goes mankind, there goes war.
>
> George Friedman, *The Next 100 Years* (2010)

Oil is our most important strategic resource, and it therefore has wide effects on global security. Nations have historically been ready for extreme measures in order to ensure their supply of oil. A significant portion of modern warfare has been at least partly because of natural resources. As the British navy decided to move from coal to oil after the First World War, they also started the era of oil as both the central fuel for waging war and a significant reason to wage war for. The armies discovered oil's usefulness in land-based warfare as Germany proved the effectiveness of armored and mechanized warfare during the Second World War. The air forces did not even exist until the era of fighter planes that ran on oil after the Second World War.

Below are a few examples of key events in world history, where oil played an important role:

- The oil embargo set up by the United States forced Japan either to submit or to escalate in 1941, which it then did by attacking Pearl Harbor.
- The leadership of Germany was forced to divert its forces in the Eastern Front from the politically and economically more central areas to the Caucasus in order to ensure its supply of oil.
- In the 1940s Britain was forced to open the Middle Eastern oil production areas to U.S. oil corporations in return for their help in the war effort.
- At the beginning of 1950s Persia (Iran) decided to nationalize its oil production. To counter, Britain declared an embargo on Persian oil exports. In 1953 Britain and the United States staged a coup, which led to Shāh Pahlavi's rise to power.

- In the 1970s the Arab countries protested about western countries' support of Israel's war effort by limiting their oil production and exports and causing widespread economic turmoil in the west.
- In the 1980s the United States precipitated the fall of the Soviet Union by raising the costs of the arms race on the one hand, and by facilitating lower oil export revenues for the Soviets on the other (for example by encouraging Saudi Arabia to increase production, which lowered oil prices).
- As the invasion of Iraq was planned in the early 2000s, it got an apt but politically inconvenient name OIL (Operation for Iraqi Liberation). After taking over Baghdad, the first secured building was the Ministry of Oil, while the priceless treasures of the national museum were left for pillaging. Lucrative rebuilding contracts had been given to U.S. corporations well before the war began.

Oil is power

The relative geopolitical power of oil exporters is bound to increase compared to the importers. This power moves especially from western countries to the Middle East and Russia.[83] On the other hand, those customers with the ability to pay for oil, such as China, will also increase their relative power. It is an intriguing question as to how these nations will wield that power. Russia has traditionally wielded international power, which it has again been gathering since the fall of the Soviet Union due to rising oil prices and export revenue. Many oil exporters have much less experience in global geopolitics, and many are also politically less stable. They also lack the massive nuclear weapons arsenal, which makes threatening or pressuring Russia a difficult task.

Oil transportation is an Achilles heel for many nations. The United States imports surprisingly little of its oil from the Middle East. But its ability to secure or block oil shipments from the area at will makes eastern Asia and Europe, for example, dependent on the goodwill of the U.S. This helps to explain the recent boom in building oil-pipeline networks from oil-producing areas in Russia and central Asia towards eastern Asia. The United States is aware that its position of power in global geopolitics is dependent on its ability to secure sufficient energy resources. In a repetitive, if rather toothless, chorus, all of the recent presidents have emphasized the need to cut U.S. dependency on imported oil. Even the premise that the quite sovereign power held by the nation with by far the largest military forces is starting to slip away from its grip, creates a breeding ground for tensions, countermeasures, and conflicts.

The living space of western people has historically been secure and calm for the past few decades.[84] Of course we get our daily dose of war, famine, violence, and unrest through television and the news media. But all of that seems to be happening somewhere else, quite far away. As tourists or on business trips, we rarely end up in the middle of a conflict. With the hyperactive 24-hour news cycle, even those remote conflicts are quickly brushed aside to give space for the upcoming election or the international relief mission. We can breathe a sigh of relief and tell ourselves that the situation is being taken care of.

The power that is associated with oil is often handled away from the public view. The participants that aim for economic development have a common interest that the oil is sold on the open market. Bigger players on the other hand get a relative advantage if they can secure their oil deliveries with long-term, bilateral contracts. These contracts involve much more than just oil and money. Security guarantees, counter-trading, commitments for involvement in developing other economic sectors, and political deals are promises of reciprocal services. For smaller oil exporters, it is important to look for bigger allies. This will help deter other less-cooperative nations and bodies from lusting after their resources.

In the event that resource scarcity gets worse, globalization slows down and the geopolitical situation shifts, the situation could take a turn for the worse. We might go back to the situation where doing well as a nation meant aggressiveness (wars and conquests included) in securing enough resources. At least from the perspective of risk management, these possibilities need to be taken into account. On the other hand, increased military preparation creates a situation where everyone else also needs to increase their military preparedness. This will demand even more of the already scarce resources from everyone.

War

For modern warfare, peak oil will mean new kinds of challenges. Modern armies are utterly dependent on oil to power them. With increasing prices and possible shortages, armies face both tactical and strategic needs for a redesign of how they operate, manage and use energy. Tactically it becomes more important to secure the availability of fuel, and strategically in the longer term the forces face a challenge to increase their energy efficiency. In the future, we will need to kill each other not just efficiently, but energy-efficiently.

The U.S. armed forces, with 77 percent of their energy coming from oil, have made consistent plans to reduce their dependence on foreign oil. These plans include for example the following:

- by 2011 to test and certify all aircraft and systems against 50/50 alternative fuel blend;
- by 2015 for the navy to reduce petroleum use in its 50,000-strong commercial fleet by half;
- by 2016, to be prepared to cost-competitively acquire 50 percent of the air force's domestic aviation fuel requirements via an alternative fuel blend in which the alternative component is derived from domestic

> sources produced in a manner that is greener than fuels produced from conventional petroleum.[85]

In Europe, there has not been much in the way of similar plans. One factor is the amount of resources available. The budget of the U.S. armed forces is comparable to the budget of the armed forces of the rest of the world combined.

The trend of military mechanization has been to increase firepower and mobility, which allows the penetration of the enemy's defenses. Manpower has been increasingly replaced with machine power. This adds to the dependence on external energy, which, save for the nuclear-powered fleets of the navies, usually means oil. In the future, military structures and military tactics that use little oil will become more attractive.

In fact the trend has been moving towards asymmetric warfare, which includes guerilla warfare and terrorism. Another trend in resistance movements and insurgencies is in the use of forces that mix in with the local population and only strike out locally and suddenly. This prevents the opposing force from gaining definite control of the area and increases their costs both in material and human lives. A model example of these tactics is Afghanistan, where local guerilla forces have worn down and expended tremendous amounts of military hardware from both the Soviet Union and for the last decade, the United States. Further, cyber-attacks, media manipulation, and various economical ways of adding pressure are examples of the fourth generation of modern warfare that blur the lines of conventional warfare.[86]

In an interview, Dr. Jarno Limnell, an expert on global threats and cyber security and military sciences, shared his views on the increasing weight of natural resources and what it means for Russian power in global geopolitics:

> As resources get scarcer and more expensive, their role in geopolitics will grow. Even today, the nations with strategic oil and gas pipelines have grown in importance, and this competition will only increase. Energy resources are the main reason why Russia is rapidly gaining global influence once again. Europe needs to pay attention as it is by no means out of the question that Russia will use oil deliveries as a tool to further its political aims in Europe.

According to Janne Malkki, who researches the history and future of warfare at Helsinki University:

> The result might be a situation, where old fashioned conscription forces increase their relative military efficiency as military technology and the oil that powers it keeps getting more expensive. The trend might also be that we will be less enthusiastic about going on expensive crisis relief missions around the world.

Some interesting points, if the threat of war started to increase, would be:

1. Around one-third of European oil comes from Russia, and another third from OPEC countries. What does this mean for the national defense forces that have grown more and more dependent on oil in their operations?
2. What would the geopolitical situation be like, and what would be the steps that led to it, if for example Russia decided to limit its oil and gas deliveries to some European countries?
3. What are the common European interests to act in for example Africa, Middle East or the Arctic region, if an increasing portion of our energy is imported, and the supposedly common energy policy gets buried under bilateral contracts and agreements?

It is quite hard to get much information out of the army or national defense forces of a country on the matter of how prepared they are and for what kinds of threats. The military might be one of the sectors that takes the subject quite seriously, but on the other hand it has very limited possibilities to actually do something about it, at least while normal times prevail. In many countries the military has been willing to start discussing the matter with other sectors of society. It remains to be seen how seriously other sectors take the challenge.

Peak oil and climate change

The acceleration of climate change has been studied and confirmed by thousands of scientific studies, as has mankind's substantial part in it. The burning of fossil fuels, which releases greenhouse gases such as carbon dioxide, is our greatest contribution to it (around two-thirds of emissions). As we add more greenhouse gases to the atmosphere, an ever-increasing amount of heat energy becomes trapped in the planet's atmosphere, oceans and soils. More of humanity's contributions include deforestation, land use changes (due to food production, for example) that release carbon from the soil, and changes in the earth's reflectivity (the albedo effect).

Will peak oil save us from runaway climate change?

With peak oil, and the prospect of peaking in other fossil fuels in the coming decades, one might think that we might not have as much to burn as the various climate change models predict. This is correct, for example according to the IPCC (2007) scenarios where we keep increasing the burning of fossil fuels but not accounting for the geological limitations in getting these fuels. This, however, is only part of the story.

By burning just the already known oil reserves, we will greatly increase the risk of passing the 2°C limit. After that, the risks for a runaway climate change grow substantially. Almost half of this warming (0.8°C) has already happened. With the slow mechanisms of the earth's climate system, the warming will continue for another 0.8°C even if we stopped all burning today. As the climate models keep getting better and more data are gathered, there are slight adjustments to the speed and severity of the climate change. From humanity's point of view, most of these adjustments are in the wrong direction. The effects of the warming climate have started to appear much faster than previously thought. One glaring example is the diminishing volume of the Arctic ice and the worrying possibility of having an ice-free Arctic during this decade.

Emissions budget

So far, we have emitted roughly 500 gigatons of carbon dioxide, and its concentration in the atmosphere has risen from 280 ppm (parts per million) to 400 ppm. The total budget researchers have given us, which allows us a chance of between 63 and 92 percent of staying below 2°C, is 886 gigatons by 2050. We have around 565 gigatons left in our budget.[87]

- If we burn our known oil reserves, we will emit around 550 gigatons.
- If we burn our known natural gas reserves (not including shale gas), we will emit around 350 gigatons.
- If we burn our known coal reserves, we will emit around 2,000 gigatons.

Depending slightly on the research and on our willingness to take risks, we have to leave somewhere between 65 and 80 percent of our known fossil fuel reserves in the ground. Anything even resembling this is not to be seen, as oil corporations and others use billions of dollars annually to search for yet more fossil fuels. When they do, the media presents the big headlines as great news. The schizophrenic conflict is so glaring that it is actually hard to realize and comprehend.

If we end up burning most of our current known reserves, the experts say that we face a warming of 4–6°C. This makes it very likely that numerous feedback loops will kick in, and we face runaway climate change. These will very likely turn the planet into a different one from that where human civilization developed.[88] [89] Food production, for example, will suffer from decreasing yields, and in some areas viable agriculture will stop altogether.

Even if our ability to burn fails to meet the predictions of most IPCC models and scenarios, carbon dioxide remains in the atmosphere for centuries. In our time frame of several decades, it does not really matter how fast we burn our fossil fuels, if we end up burning them all anyway. Positive feedback loops will kick in without much warning, after which there is little we can do about it, emissions or no emissions. With peak oil and the host of associated risks, problems and crises that it potentially brings, mitigating climate change can be all but forgotten. It takes all the available resources we have just to cope with the short-term problems, so long-term problems get buried out of sight.

The situation is made gloomier with aerosols. These small particles that the burning of dirty fuels produces, have reportedly cooled the planet by several tenths of a degree, as they reflect part of the sunlight back into space. But unlike carbon dioxide, these aerosols only remain in the atmosphere for some weeks or months. If we cut our burning, we will face a rapid warming as the cooling effect of the aerosols will diminish.

Could the economic problems and recessions bring consumption down permanently, and so stop the climate from warming? First there is no way to stop the warming, as it takes possibly millennia for the oceans to cool to their previous levels even if we stopped emitting today. Second, it is hard, if not impossible, to cut our use of fossil fuels, especially oil, as societies and humans currently have no choice but to use them. The mechanism to stop burning simply does not exist. It is true that as prices rise, demand will be destroyed, especially in the developed world that uses much more than its fair share. But the developing countries would most likely use those resources. And as demand decreases, so does the price, and pretty soon someone would be able to use the fuels for something. As emissions would decrease due to recessions, so would our ability and willingness to invest in low-emissions energy production. Only a decrease in the global amount of emissions will help slow down climate change. Local, or even continental, temporary economic fluctuation is not enough.[90]

So what would this carbon budget mean for the average western citizen's life? The per capita emissions between OECD countries vary greatly, but let's assume emissions of around 10 tons per year per person. This can be compared with the per capita emission of 1 ton in India. A westerner needs to cut emissions by 8 tons, and bring them down to 2 tons per year. This would mean that we could drive around in our cars and move pretty much normally. And nothing else. No heating or cooling. No electricity. No shopping, no food. No work. No services. It is mind-boggling how the actual massive size of the emission cuts needed has not dawned on politicians, citizens or even most of the environmental groups, no matter how many scientific reports we have published. We seem to think that we actually can afford to oppose almost all of the low-emission options to produce our energy, from wind power to nuclear, with one excuse or another.

Energy security, peak oil, climate change . . .

Mitigating peak oil will also mitigate climate change, but only if we concentrate on making society less dependent on oil, instead of concentrating on finding and securing access to more and more substitutes. If we fail to make a controlled transition away from oil dependency in time, we will have to react to oil shortages by any means necessary, including burning any substitutes we can find.

The problem is that dealing with peak oil in the short and medium term endangers any efforts to mitigate climate change or to reach any meaningful global climate treaty. Reacting to peak oil in the shorter term often includes:

- bilateral, long-term contracts on oil and other key commodities;
- increased efforts to find and produce more oil due to its rising prices;
- development of more marginal unconventional oil;
- incentives to build more CtL, GtL and BtL plants, that will add to total emissions.

We have already seen the first impacts of the collision between energy security, economy and climate change mitigation. The United States, and recently Canada, have been reluctant to agree on any emissions targets, as Canada gets significant income from its oil sands production. Meanwhile, the same nations aim to increase their car fleets' fuel efficiency, which would help with climate change mitigation. In 2008 President Barack Obama said that he would start healing the planet. In 2012 he boasted that the United States had a record number of oil rigs operating – more than the rest of the planet combined. China is the fastest-growing nation for new nuclear, wind and solar power, but it also burns almost half of the global coal with new coal power plants coming online almost weekly. There is a host of unsolved conflicts between economic growth, mitigating climate change, and increasing energy security.

There is a fundamental problem with most of the climate actions we are taking. Reducing car emissions can seem like a good idea, but the only thing that matters from the actual physics point of view is total emissions. So if we end up driving more with the more efficient cars, we have not really accomplished anything, but perhaps increased congestion. Measuring emissions per unit of economic growth seems intuitive, but it obscures the fact that actually we would need to measure and concentrate on total emissions much more.

Cutting emissions and peak oil

How does mitigating climate change compare with mitigating peak oil? With climate change, the actions are often taken looking at a few decades ahead. The consequences of peak oil on the other hand are usually seen almost immediately, and the actions taken must be fast. The speed of the change and crises are very different. The end of cheap oil and slow growth of expensive oil are wreaking havoc on the economy as we speak, even though technically speaking the actual peak in all liquid fuels production may be years ahead. The scale of the sudden economic and social consequences of the peak might be comparable, from the perspective of modern human civilization, to those caused by climate change in the longer term. Further, it takes at least 10–20 years of real effort to restructure society and infrastructure to be much less oil-dependent.[91] This kind of effort is yet to be seen anywhere. An SUV sold now will be in use far into the 2020s. A coal plant built now will be operational until at least 2050. According to the IEA, after 2017 all plants and factories must be zero emissions, as by then we will have locked in enough emissions with the current infrastructure to reach 2°C of warming.

It is simply astounding how mankind is still largely debating if these phenomena actually exist at all, as the last logical day of massive mitigation efforts is far in the past. This applies for both peak oil and climate change.

Because of rising energy prices and economic problems, peak oil can cripple any sufficient long-term investments needed to mitigate climate change. We seem to be capable of long-term environmental protection only when there is economic growth and rising living standards. Prosperity (or survival) in the short term almost always wins against long-term goals. According to a memorable quote by Professor Jukka Tuomela, "It seems that we can only save earth if there is profit in it."

+2°C

The "safe" limit of 2°C of warming, and the limit of 450 ppm of carbon dioxide in our atmosphere, has been our common aim only because we have agreed on it. Both +2°C and 450 ppm are a game of Russian roulette that we play, and the stake is our future. Different countries have stakes of various sizes, as the effects of climate change will be distributed unevenly. According to some scientists, such as James Hansen from NASA, 450 ppm is too high, and he has put the safe longer-term limit at 350 ppm. That level was surpassed in the last century. This might seem unfair for many, since many of us were just getting used to the knowledge that something needed to be done about our emissions at some point, and that science and technology would have plenty of time to figure out how to do it in a non-invasive way. This situation resembles peak oil. A few years ago, with undeniable evidence, the peak of conventional oil production was transferred from far in the future to a few years ago in the past. Along with this realization, most hope of a prepared, calm and orderly mitigation was flushed down the toilet. Nature and the laws of physics do not care what seems nice or fair to us, and neither do they negotiate. The truth is not in the middle of opposing extremist views, but where it is according to the laws of physics.

One might imagine that humanity's emissions reductions would, after decades of knowledge and negotiations, be well on their way. The reality is quite the opposite. We have barely been able to slow the growth rate of annual emissions, let alone actually reduce the level of emissions. Now we face the need to cut even more emissions even faster, as the global economy is going from one crisis to the next. We simply do not have time to wait for the next breakthrough that may or may not be put together in some laboratory. Either we take all the solutions that are on the shelf and apply them with almost unprecedented determination right now, or we have lost. Many environmental groups and green politicians have cheered Germany's decision to ramp down their nuclear power stations. It is an ultimate irony that Germany is building more and more coal-fired power plants to take the place of nuclear power plants, and their emissions are growing when they should be dropping fast. These coal plants will be operational for up to half a century. The touted redemption for fossil fuels, especially coal, called carbon capture and sequestration (CCS), is an unproven technology on the necessary scale. It has mainly been used to help pump more oil, which justifies the expenses. CCS for coal simply makes coal power less efficient and more costly, as large amounts of energy are needed for the capturing, transporting and sequestering of the carbon dioxide. Since it takes energy, it actually increases the pressure to dig for more coal. A massive and swift roll-out of CCS however has a major role in the plans of the IEA, IPCC, and OECD to cut emissions in the coming decades.

What does it mean if we lose? It means self-enforcing climate change that we can do little about. With peak oil, it means crumbling economies, increasing geopolitical problems, resource wars and grabbing for natural resources. It could mean the end of globalization as we know it, and nobody knows what comes after that.

As we approach the turning point where climate change starts to feed on itself irreversibly, we still have the possibility to mitigate the situation with geoengineering. Geoengineering is a toolbox that includes a range of ideas from reasonably safe to really wild ideas, with various degrees of effect, costs, risks and side effects. A collection of these tools can be found in Risto Isomäki's "66 Ways to Absorb Carbon and Improve the Earth's Reflectivity."[92] The ways include anything from painting the roofs of buildings white to spreading sulfur particles in the atmosphere to reflect sunlight.

While planning for these methods might be a good idea, it is also seen as an excuse to continue burning. In addition, increasing the reflectivity of the planet does nothing for the problem of acidifying oceans. This is threatening many ocean ecosystems and species, and of course fisheries which we depend on for food. Most of the ways also need constant maintenance, so we would have to continue doing them for centuries. If the artificially increased reflectivity were to suddenly drop at some point, we would face a very sudden warming. Sudden changes are the ones most harmful for different species, as evolution would not have time to adapt. There are many nations capable of geoengineering. There has been urgent pressure to agree on global rules on what should and should not be done, as the actions of one nation affects all the others.

International cooperation

Mitigating and slowing climate change requires consistent, international efforts and firm limits on emissions. This is hard to mix with the added pressure of nations to increase their energy security for their own interests with the unfolding of peak oil. Uncontrolled peak oil can therefore start a chain of events that will lead to our grandchildren experiencing uncontrolled climate change as well. At that point, they may not have sufficient energy and resources to adapt to and to protect themselves from the effects of a more volatile climate. A more controlled peak oil does not stop this chain of events, but at least it gives us a chance to do something.

A positive example of international cooperation is the effort to stop the destruction of the ozone layer. The ozone in the upper atmosphere protects us and our planet from deadly UV-radiation, and the first science on the matter appeared in 1976. In 1985 the world was shocked by the news that the cause of the significant reduction in the ozone layer above the Antarctic was human-induced chemicals such as CFCs. The future seemed scary, but as early as 1987 the industrialized nations signed the Montreal protocol that severely limited the use and release of CFCs. These emissions have been plummeting, and the ozone layer has started recuperating, though it will possibly take half a century.

Around the same time as the Montreal Protocol, the Brundtland Commission published its report on sustainable development, and the road was open for the climate meeting of 1992 in Rio de Janeiro (UNFCCC). After such a brisk start, how have we managed to end up at the dead end in which we now find ourselves? One reason is that CFCs are not the same thing as a major source of primary energy for society. Removing CFCs does not hit economic growth, society's energy security, and our ability to function. A replacement for CFCs was found relatively easily. It seems that industrial nations are able to make quick adjustments and cuts as long as they do not hurt economic growth. With replacing fossil fuels, the situation seems quite different, and there seems to be no hurry. Despite decades of negotiations, promises and hype, the promise of significant green growth has failed to materialize, along with global emission limits and carbon taxes. It seems that we cannot both eat the cake of economic growth and save its benefits for future generations to enjoy.

Summary – How will oil consumption shrink and from where?

It is important to realize that there will always be as much oil as is consumed. Or conversely, the amount of oil that can be consumed is the amount that is available. Price, along with a few other factors will determine who will use the available oil and how. The basic assumption of the market is that the oil will be used by whoever can pay more for it than the others. If there is a single barrel of oil available with someone able to pay 99 dollars and someone else able to pay 100 dollars, the latter will get the oil. They can use this oil as they see fit. If they use it in a productive manner, meaning that they create over 100 dollars worth of value (goods or services), then it might be that they could afford to buy their oil also in the future. If they use it for cruising around in a sports car, it might be that the next time they will not be able to afford as high a price, and the barrel of oil will end up somewhere else.

In short, if there is a bloodthirsty pack of wolves running after a group of people, you only need to run slightly faster than the slowest person does. The actual speed does not matter that much.

Other factors to count are capital, for example various assets, the rents and profits collected from these, and the resulting ability to borrow against these assets. To continue the example above, the one with the sports car cruising around can use their fancy car to borrow against. That way he would be able to continue cruising, at least for a while. The rich proportion of the population who own capital assets can also finance their sports car hobby with the profits from those assets. Often the amount of capital, and therefore the profits, is quite large (and the amount of people in that group rather small). This means that there would be enough oil for all of them, no matter what they did with it or how much they used. This group has recently been labeled as "the 1 percent" (leaving the rest as the 99 percent). The increased costs of oil are invisible to this group, as they represent such a small part of their budget. It is regrettable that it is this same group of people who also control national and international politics and make the decisions in corporations. A quote from a top politician will help to give a glimpse of this alternative reality: "Peak oil should not be a significant problem, since the new Audi A3 consumes much less gasoline than the older models and is still as powerful."[93]

According to the mechanisms described above, the market will destroy demand in slightly different places depending on the time frame. In the short term, the victims will be those who lack liquidity at the time. With a longer time frame, the effects start to spread upwards in society to those who have some liquidity and who are able to borrow a limited amount of money, namely the lower middle class.

When one looks at the decrease of oil consumption on a large scale, for example on a national scale, the decrease always seems to be gradual. The exception is rapid societal collapse as seen in several euro-countries recently, where oil consumption has crashed very fast. It seems that this statistical phenomenon is behind the views of many economists (and others), who keep insisting that people will gradually drive less when oil becomes more expensive. It might well be an illusion where one does not see the important details because of the whole.

With individuals and inside smaller groups, the change in oil consumption is often sudden. Weekend trips to the cottage or the amusement park will be made, since the gradually increasing prices of fuel never seem to cut too deep in the pocketbook, which would cause the family to stay in the city apartment. Commuting to and from work and hobbies are also done normally no matter the price of gasoline. The purchasing power lost due to more expensive gasoline is taken from somewhere else. There will be enough money for gasoline, until, for example, unemployment dramatically cuts consumption and the ability to drive to work or to former expensive hobbies. It seems that gasoline offers such a high degree of mobility and freedom to many people that they are not willing to significantly cut driving unless they absolutely have to.

Part of the reduced oil consumption is hard to recognize. It is hidden in production chains, the costs of subcontractors, and the prices of imported goods. As we buy fewer goods and services, the hidden oil embedded in them is also reduced. In services the amount of this oil varies, but is usually rather small. Producing services might offer an oil efficient way to produce economic activity. This is the mechanism that the western nations have seen in recent years, as their economies have appeared to decouple from their oil use. It should be noted that most services are voluntary, so they will get cut first when purchasing power decreases. This can shift the recent decoupling of oil and economic growth into reverse, causing the economy to shrink more in relation to oil use.

To summarize, inside nations oil demand is first reduced by the poor, from where it will spread towards the middle class, making them poorer. As the economy continues to deteriorate, non-essential services and businesses will generally be hit the hardest. The mechanism in itself increases inequality in society, and in time it will start to choke the middle class as well, as many workers in service sectors face unemployment. Even if a job is found somewhere, it could mean increases in commuting costs, which will decrease purchasing power.

This market mechanism is distorted by many factors. Social security offers the unemployed some kind of possibility to continue consuming. One can also count the many gasoline subsidies in oil-producing nations as a form of social security. From society's point of view, it would likely be sensible to offer essential mobility such as commuting for work and services, for as many people as possible. In a tightening situation this adds pressure to reduce less productive use, such as leisure use.

The effects of sudden supply crunches or price increases are also mitigated by the emergency reserves held by IEA members. Even the hint of releasing oil from these supplies is enough to calm the markets, at least temporarily. Many countries are aiming to increase the use of natural gas as an energy source. Despite some talk, there has been little in the way of building a comparable emergency supply for natural gas.

Below are few conditions that negatively affect a nation's ability to pay for its oil. For example in many eurozone countries that have found themselves in crisis, almost all of these conditions hold true, so other countries will have to run just slightly faster than these countries. Due to the interconnectedness of global finance and trade, the problems in a few nations will eventually spread elsewhere.

1 Countries that lack their own oil production and see the direct effects of rising oil prices in their trade balance, whereas the effect will be the opposite for oil exporters.
2 Countries that use oil inefficiently, for example in transportation or electricity production.
3 Countries that are deep in debt, and therefore have less chance to get more debt than others, resulting also in a bigger share of the national income going to servicing that debt.
4 Countries that base their economy on sectors that are sensitive to oil prices and economic fluctuation; for example mass tourism that is dependent on a wealthy and large middle class and cheap flights.
5 Countries that lack a competitive exports industry. Imported oil is paid for by exporting goods or services to other countries or, in the case of tourism, by attracting foreign consumers and their money to one's own country.

The last chapter of this book ties up some of the loose ends and offers some thoughts and perspectives on how peak oil or resource scarcity may be observed in society. What can we do to mitigate the situation, assuming that we are stuck with our current society and infrastructure? What does the world look like after cheap oil?

Notes

1 IEA *World Energy Outlook* 2013, p. 511.
2 http://energyforumonline.com/date/2013/03/page/4.
3 IEA *World Energy Outlook* 2011, p. 109.
4 http://ec.europa.eu/energy/observatory/countries/doc/key_figures.pdf.
5 BP *Statistical Review of World Energy* 2014.
6 www.ft.com/cms/s/0/01ba1a04-2c24-11e3-acf4-00144feab7de.html.
7 www.eia.gov/forecasts/steo/index.cfm?src=Petroleum-f1.
8 www.eia.gov/tools/faqs/faq.cfm?id=727&t=6.
9 *Monthly Energy Review*, July 2013, EIA, www.eia.gov/totalenergy/data/monthly.
10 http://deepblue.lib.umich.edu/bitstream/handle/2027.42/99124/102951.pdf?sequence=1.
11 http://ourfiniteworld.com/2013/01/31/why-is-us-oil-consumption-lower-better-gasoline-mileage.
12 *Helsingin Sanomat* 30th July 2006 (in Finnish).
13 www.bbc.co.uk/news/world-europe-25460427.
14 www.zacks.com/stock/news/67191/BP-Rosneft-Fall-Out-of-Sakhalin-5.
15 http://english.ruvr.ru/news/2013_07_15/Russia-revealed-extent-of-oil-and-gas-reserves-for-first-time-1435.
16 http://eng.transneft.ru/news/509.
17 www.eia.gov/countries/cab.cfm?fips=RS.

18 Aleklett, K. (2012), *Peeking at Peak Oil*, Springer, p. 283.
19 www.inference.phy.cam.ac.uk/withouthotair/c20/page_128.shtml.
20 MacKay, D. (2008), *Sustainable Energy – Without the Hot Air*, UIT, www.withouthotair.com.
21 www.msnbc.msn.com/id/20095291/ns/business-eye_on_the_economy/t/us-highway-system-badly-need-repair.
22 www.irfnet.eu/index.php?option=com_content&view=article&id=261&Itemid=61.
23 http://online.wsj.com/article/BT-CO-20130625-706882.html.
24 *Helsingin Sanomat* 23 February 2013 (in Finnish).
25 www.theicct.org/sites/default/files/publications/Pocketbook_LowRes_withNotes-1.pdf.
26 www.autonews.com/apps/pbcs.dll/article?AID=/20130828/RETAIL01/130829913/u-s-sales-pace-projected-to-top-15-5m-in-august#axzz2dGBoULHP.
27 www.theicct.org/sites/default/files/publications/ICCT_New_Passenger_Car_Fleet_China_2010.pdf.
28 www.iea.org/newsroomandevents/pressreleases/2012/september/name,31383,en.html.
29 Blees, T. (2008), *Prescription for the Planet: The Painless Remedy for Our Energy & Environmental Crises*, BookSurge Publishing.
30 www.hurriyetdailynews.com/default.aspx?pageid=438&n=turkey-tests-its-first-boron-fuel-cell-car-2011-11-07.
31 Aleklett, K. (2012), *Peeking at Peak Oil*, Springer, p. 134.
32 http://science.howstuffworks.com/transport/flight/modern/question192.htm.
33 Tamminen, T. (2008), *Lives Per Gallon: The True Cost of Our Oil Addiction*, Shearwater.
34 www.trafi.fi/palvelut/julkaisut/2012_julkaisut/suomenlentoliikenne (in Finnish).
35 Mair, S. (2011), Piracy and maritime security, www.swp-berlin.org/fileadmin/contents/products/research_papers/2011_RP03_mrs_ks.pdf.
36 www.eia.gov/cabs/world_oil_transit_chokepoints/full.html.
37 Gilbers, R., Perl, A. (2010), *Transport Revolutions*, New Society Publishers, pp. 97–99.
38 http://en.wikipedia.org/wiki/Rail_transport_in_Europe.
39 http://decarbonisesa.com/2013/01/10/bounding-the-future-essential-thinking-on-population.
40 www.scientificamerican.com/article.cfm?id=human-population-growth-creeps-back-up.
41 Czech, B. (2013), *Supply Shock: Economic Growth at the Crossroads and the Steady State Solution*, New Society Publishers.
42 www.fromthewilderness.com/free/ww3/100303_eating_oil.html.
43 PEAK OIL – Security policy implications of scarce resources (2010), Bundeswehr Transformation Centre, Future Analysis Branch, chapter 3.1.4
44 Norman Borlaug was one of the central scientists who worked on plant breeding. More can be read on him in the book *Our Daily Bread; The Essential Norman Borlaug* (2011) by Noel Vietmeyer.
45 Tainter, J., Patzek, W. (2011), *Drilling Down*, Springer-Verlag New York Inc.
46 Brown, L. (2012), *Full Planet, Empty Plates – The New Geopolitics of Food Scarcity*, W.W. Norton & Co.
47 http://en.wikipedia.org/wiki/Erosion#Land_degradation.
48 Smith, K., Edwards, R. (2008), 2008: The year of global food crisis, *The Herald* (Scotland), 8 March 2008, www.heraldscotland.com/2008-the-year-of-global-food-crisis-1.828546.
49 unctad.org/en/PublicationsLibrary/ditcted2012d3_en.pdf_.

50 Brown, L. (2012), *Full Planet, Empty Plates – The New Geopolitics of Food Scarcity*, W.W. Norton & Co, p. 102.
51 http://epp.eurostat.ec.europa.eu/cache/ITY_OFFPUB/KS-FK-12-001/EN/KS-FK-12-001-EN.PDF.
52 http://epp.eurostat.ec.europa.eu/statistics_explained/index.php/Agricultural_census_2010_-_main_results.
53 http://ec.europa.eu/agriculture/agrista/tradestats/index_sem.htm.
54 http://energyefficiency-fisheries.jrc.ec.europa.eu.
55 www.ers.usda.gov/topics/international-markets-trade/us-agricultural-trade/export-share-of-production.aspx#.UtO0m1tdVSI.
56 http://en.wikipedia.org/wiki/Agriculture_in_the_United_States.
57 www.census.gov/compendia/statab/cats/agriculture.html.
58 www.geneticliteracyproject.org/wp/wp-content/uploads/2013/10/Nicolia-20131.pdf.
59 http://en.wikipedia.org/wiki/Golden_rice.
60 www.geneticliteracyproject.org/wp/wp-content/uploads/2013/10/Nicolia-20131.pdf.
61 www.easac.eu/fileadmin/Reports/Planting_the_Future/EASAC_Planting_the_Future_LAY_SUMMARY.pdf.
62 www.aaas.org/news/statement-aaas-board-directorslabeling-genetically-modified-foods.
63 http://ec.europa.eu/research/biosociety/pdf/a_decade_of_eufunded_gmo_research.pdf.
64 www.nap.edu/openbook.php?record_id=10977&page=8.
65 www.who.int/foodsafety/publications/biotech/biotech_en.pdf.
66 www.imf.org/external/pubs/ft/wp/2012/wp12109.pdf.
67 www.oecd-ilibrary.org/economics/the-price-of-oil-will-it-start-rising-again_5k49q186vxnp-en.
68 http://ec.europa.eu/energy/observatory/countries/doc/key_figures.pdf.
69 Jylhä, P. (2011), Harvesting undelimbed Scots pine (*Pinus sylvestris* L.) from first thinnings for integrated production of kraft pulp and energy. *Dissertationes Forestales*, 133. www.metla.fi/dissertationes/df133.htm.
70 Mair, S. (2011), Piracy and Maritime Security, www.swp-berlin.org/fileadmin/contents/products/research_papers/2011_RP03_mrs_ks.pdf.
71 www.eia.gov/cabs/world_oil_transit_chokepoints/full.html.
72 www.worldenergy.org/documents/congresspapers/365.pdf.
73 www.worldenergy.org/documents/congresspapers/365.pdf.
74 www.eia.gov/dnav/pet/hist/LeafHandler.ashx?n=PET&s=RBRTE&f=A.
75 Ayres, R., Warr, B. (2009), *The Economic Growth Engine: How Energy and Work Drive Material Prosperity*, Edward Elgar.
76 ITPOES 2010: Dr. Robert Falkner, LSE Dept. of International Relations, p. 30.
77 Ayres, R., Warr, B. (2009), *The Economic Growth Engine: How Energy and Work Drive Material Prosperity*, Edward Elgar.
78 www.iaee.org/en/publications/newsletterdl.aspx?id=156.
79 http://oilprice.com/Energy/Oil-Prices/How-Closely-are-Oil-Prices-Tied-to-Economic-Activity.html.
80 Rubin, J. (2012), *The End of Growth*, Random House, Canada.
81 www.thecuttingedgenews.com/index.php?article=11440.
82 Read more on this matter in Dambisa Moyo's book *Winner Take All: China's Race for Resources and What It Means for the World* (2012) or Stephen Leeb and Gregory Dorsey's book *Red Alert: How China's Growing Prosperity Threatens the American Way of Life* (2011).

83 PEAK OIL – Security policy implications of scarce resources, (2010), Bundeswehr Transformation Centre, Future Analysis Branch.
84 More on the historical trends of violence and war and their possible reasons can be read in, for example, Steven Pinker's *The Better Angels of Our Nature: Why Violence Has Declined* (2011, Penguin Books).
85 www.cnas.org/files/documents/publications/CNAS_Fueling the Future Force_NaglParthemore.pdf.
86 http://en.wikipedia.org/wiki/Fourth-generation_warfare.
87 Greenhouse-gas emission targets for limiting global warming to 2 C, *Nature*, 458, 30th April 2009.
88 www.rollingstone.com/politics/news/global-warmings-terrifying-new-math-20120719.
89 Aleklett, K. (2012), *Peeking at Peak Oil*, Springer, p. 236.
90 www.noaanews.noaa.gov/stories2009/20090126_climate.html.
91 Hirsch, R. (2005), *Peaking of World Oil Production: Impacts, Mitigation, and Risk Management*, Department of Energy (USA).
92 www.intokustannus.fi/kirja/66_ways.
93 A top Finnish politician, Mauri Pekkarinen, from the central party "Keskusta," said this in the discussion panel held at the publication event of this book's Finnish version Suomi öljyn jälkeen. Mr. Pekkarinen was a key figure in forming the national Energy and Climate Strategy of 2008.

5 World after cheap oil

What then?

What does all this mean for you and me? In this last chapter, we aim to make some sense of it and connect some of the dots that we have presented earlier in the book. If you have concluded that peak oil and its dynamics are a potential problem, you have come a long way. With the information given in this book and elsewhere, you can also draw your own conclusions on what all of this means for you. How will you react, if you do? You are the only one who can think for yourself, draw conclusions and see how things and actions coexist and affect one other. It is probably not wise to believe everything that the business-as-usual crowd is saying. The same probably goes for those predicting impending doom, destruction and the swift end of the world as we know it. Then again, if even some of the trends and risks mentioned in this book actually happen, is it wise not to prepare at all? In risk management, the central topics are the likelihood of risk, the severity of its consequences, and the costs of mitigating and preparing for it. When it comes to peak oil, the first two seem to be largely underestimated, and the mitigation part largely ignored.

Peak oil, if the officials and politicians are even aware of the concept, is just one more problem on their list of problems. The decision makers' time is spread thinly among many competing topics and special interests. It would seem that in many countries no one in the political system is actually responsible for following the oil topic on a more holistic basis. And if no one is following the subject, the risks it contains go largely unnoticed. And as they remain unnoticed, nobody sees any urgency. Someone has to take the first step.

Dealing with a challenging topic such as peak oil on an individual and societal level requires many steps. Psychology recognizes many mechanisms that make processes like this harder and slower. Cognitive dissonance, resistance to change, confirmation bias, trust in authorities and groupthink are just a few. You can read more on these at Appendix 3 "Reason alone won't make change happen."

Each of us deals with these processes at our own pace. Often the first reaction to higher oil (and therefore gasoline) prices is the demand to lower the taxes on gasoline and car ownership. This would make it possible to continue as before for a while longer. Behavior often changes only when it becomes nearly impossible to continue old habits. This can happen due to legislation or costs, for example. A more systemic shift occurs when masses change their behavior. Reaching a tipping point like this requires that the masses get new information and start changing their attitude in the light of it. These kinds of changes are hard to predict and even harder to control. It is therefore not in the interests of the current power structures to cause a fast shift in mass psychology. Actually, it is not in the interests of most of us, given the unpredictability of possible events. Most of us will have to face these changes whether we want to or not. The question is mainly about the time frame and our abilities and tools at the moment.

As the scheme of our debt-driven consumerism upon which we are building our future starts to unfold, the social contract current citizens are committed to may begin to unravel. In addition to economic inequality, we might see tensions building between other groups as well. These can be different generations, rural and urban populations, public and private employees, tax paying and non-taxpaying citizens, or different business sectors that are more or less crucial when it comes to oil (more on these later). At the societal and national levels, these kinds of shifts are usually accompanied by hardening attitudes. The voices of the different special interest groups can only grow louder.

We need to change the way we think

The world is an extremely complex place that is notoriously hard to predict. This is not an excuse not to do anything, as we know enough that we can start in the right general direction. All actions that lessen our dependency on oil are important, but they (as well as their side effects) need to be evaluated before they are implemented. We have to find and use different substitutes for oil, as we are terribly late with mitigation and preparation. However, it would be utter madness to trust these substitutes in the longer run and continue as if there is no problem at all. The way we measure the viability of current investments has to change.

The dominant thinking for the last decades, until a few years ago, has been that cheap oil will be available from here to infinity. Climate change and its mitigation gave birth to the topic of doing something to lessen our oil use to cut emissions at some point. Now we need to add the reality of rising and more volatile oil prices to our frames of thought. This is not to say anything about the growing possibility and potential severity that oil crises and supply shocks would have for the importing nations.

Bilateral oil trade will continue to grow, and many developing nations can outbid the developed world on oil prices. For example the PIIGS[1] countries worst affected by the euro crises are also the nations that use the most oil in their energy mix. The average oil use of OECD countries has been dropping 2 percent annually for the last few years. Part of this can be explained with better efficiency, but most is explained by the economic difficulties. From another perspective, part of the economic problem can be explained by the inability to buy and use more oil.

If all goes well, the bumpy plateau of oil production will continue for several more years, maybe even for a decade. After that, the production will go into terminal decline. The steepness and steadiness of the decline is hard to predict. With luck, it will be relatively steady and gentle. However, we have not even prepared for this. What will happen if some of the gloomier, yet relatively likely, scenarios play out? If the Arctic area fails to deliver the estimated amounts of oil and gas, if shale energy proves to be a larger bubble than estimated, or the unconfirmed political reserves in OPEC countries prove to be partly non-existent? Or if Russia fails to invest as much as it needs into its energy infrastructure? Or if the planned, phenomenal increase in Iraqi oil production fails to materialize because of growing local unrest? Even if just one or two of these things were to happen, we could face declines of several percent annually in oil production.

An increasing number of respectable publications are researching and trying to model the effects of oil production volumes on the global economy. According to many of these, the economy can grow slightly even if the total oil production does not grow. But only slightly. This would not be enough for the debt-driven economy to stay viable. When the production starts its decline, it can be caused by a global recession and lack of investments in exploration and production. Or the global recession can be caused by the decrease in oil production. These phenomena feed on each other, and either of the two can be the first step.

Europeans should think long and hard about how to deal with the situation as the population gets older and domestic consumption decreases along with falling export revenues. It has been shown in research that people are willing to lower their standard of living, but only if everybody else also lowers theirs. Growing the gap between different standards of living while there is less to share is a recipe for social unrest. It also leads to less than optimal use of the nation's resources. This will eventually also hurt the rich.[2] If one looks at the situation through the lens of traditional party politics, the right-wing politicians need to understand this dynamic. The left-wing, on the other hand, need to realize that the welfare state and its social services are created primarily with cheap fossil fuels, which may not be available in the future. The system has been upheld with debt in many countries for years, whether one measures that debt in money or natural resources.

Even the richer European nations face severe problems. With expensive oil and declining tourism, the current euro crisis countries might face increasing difficulties in servicing their debts. This can mean that the rich countries will have to pick up the bill. As they will also face worsening trade balances due to rising import costs, their economies will be hit simultaneously from two sides.

Snake oil salesmen

There is no silver bullet, no snake oil, to help us easily get rid of our predicament. However, this does not mean that such snake oil will not be offered to us. Quite the contrary. We are offered misleading and reality-distorting information on a daily basis from a multitude of sources. The central question one needs to ask when presented with such information is who profits? *Cui bono*? Many owe their existential justification to the current situation and its continuance. Many others are simply selling their own solution. Not that there is anything wrong with selling one's own solution – we truly need good ideas and solutions – but the message always needs to be viewed with healthy skepticism.

We face an unprecedented situation. We must substitute our primary energy source, which is in rapid decline, with substitutes that are of poorer quality in most ways. The thing we are dependent on is *cheap* petroleum. As such, it cannot be replaced with expensive petroleum. So far, humanity has harnessed energy sources of higher and higher quality and energy density, but we have never moved away from the ones we used before. We burn more coal today than we have ever burned in humanity's history, even if the relative amount has decreased. We harvest a bigger part of the planet's capacity to produce biomass than ever before. Coal did not replace firewood, even though it probably saved what was left of Europe's forest at the time. Oil did not replace coal, and neither did natural gas. They just added to our total primary energy consumption. There are a few examples of fuel replacements in energy production. After the oil and energy crisis in the 1970s, oil was largely replaced by nuclear power (and coal) in electricity production. It took more than a decade to get back to the 1979 level of oil consumption. Modern oil production and refining also saved the sperm whale from extinction, as kerosene replaced whale oil in lighting appliances near the end of the nineteenth century.

What does peak oil look like?

> Get past the fantasies of sudden collapse on the one hand, and the fantasies of limitless progress on the other, and what you get is what we're getting—a long ragged slope of rising energy prices, economic contraction, and political failure, punctuated with a crisis here, a local or regional catastrophe there, a war somewhere else—all against a backdrop of disintegrating infrastructure, declining living standards, decreasing access to health care and similar services, and the like, which of course has been happening here in the United States for some years already.
>
> John M. Greer

Like the situation with oil, the effects of peak oil are spread everywhere in modern society. They are rarely staring one in the face, so one could point to them and say that this is because of peak oil. Oil, or peak oil, is rarely the only clear reason for a certain problem or situation. But a closer look often reveals that it is in the background of many events. One can start to grasp the role of oil with the following thought game:

1 First, imagine a situation where oil would be essentially free, and there would be enough for everyone. It would also have no hidden or externalized costs such as pollution, environmental destruction, or greenhouse gas emissions. It is easy to see that this endless, nearly free energy source would propel the global economy into a robust period of growth, at least for a time, since production of most goods and services would be much cheaper. This would lead to more consumption, which would lead to more jobs, more purchasing power and so on.
2 Then, imagine a situation where, as of tomorrow, there would not be a drop of oil available. Strategic emergency reserves would be empty in a few months. Food production, construction, and nearly all of transportation would cease. Power plants producing electricity would be taken down when their fuel storage runs dry. Humanity would rapidly sink into a situation where observing economic trends, or even staying relatively civilized, would be replaced by the single-minded drive to live another day.

It is hard for people to observe the effects of peak oil, as the time span is quite long. Liquid fuel prices keep climbing cent by cent, year after year. The economy slowly gets worse and worse, with brief periods of relief that manage to restore everyone's hopes that the growth and good times will return. Bigger collapses or shocks usually happen somewhere else. And when they do hit, it is too late to contemplate their deeper origins. It is impossible to prepare for something after it has happened.

Oil prices shot up sharply during the Arab Spring and the civil war in Libya. This was largely because there was not much spare capacity of oil production. Even if we manage to avoid and manage crises like these, and societies could continue along their present course, we will eventually face chronic supply disruptions and rationing. For the average middle-class citizen even the thought seems far-fetched. It is, however, already normal in many places, as the price of fuel cuts the poor from the oil-consuming community. The rationing is done with price, and since the poor cannot afford fuel, they also cannot afford an electric car or solar panels. With chronic supply disruptions and rationing (via price or some other mechanism), the demand is often destroyed quite permanently. This means that the standard of living collapses to a lower level, often against people's wishes. The consequences for society as a whole are rarely positive.

As cheap oil vanishes, business-as-usual is leading humanity towards a long, bumpy downhill ride in all ways; economically, ecologically, and socially. The bumps represent the ever-increasing number of problems, as our current economic and social structures try to deal with the new situation. These problems will keep the decision makers and analysts busy. The euro has bounced from one crisis to the next for years, offering the media, citizens and politicians more than enough to talk and fuss about. Who has the motivation, time, or even ability to concentrate on problems happening further down the road?

Three levels of the economy

> Growth in the real economy ceased quite some years ago
>
> Tim Morgan (2013)[3]

Economic activity can be roughly divided into three levels. There is the primary economy, which concentrates on producing raw materials and other primary products. Growing the primary economy is becoming harder and harder, as the quality of non-renewable resources keeps getting worse and even the renewables face problems as we have ended up eating the principle instead of the yield. The problem is further aggravated by the slow growth in oil production, as the primary economy uses oil as its main energy source. The secondary economy includes refining, manufacturing, logistics and various services. It is also dependent on growing energy consumption, even if in a little less straightforward way. If there is no growth in raw materials and energy, there will not be much growth in manufacturing or transportation. There will also be less need for people to travel to work and to buy and consume those goods and services.

The tertiary economy includes the sectors that are almost totally removed from any physical economy, but the border is hard to draw exactly. The tertiary economy includes most of the financial sector with their derivatives and such. They concentrate mainly on speculating and moving claims on wealth around instead of producing any actual wealth by themselves. They are not limited as such by the physical limits of the real world, and so it has been more than tempting to try to deal with economic problems by letting the tertiary industry grow bigger and bigger. The tertiary economy, finance industry included, is tightly linked with the real economy, as it offers essential services such as loans for investments.

There have lately been attempts to get the economy back on the growth track by adding debt and printing money. The total theoretical value of derivatives and other investment vehicles is estimated at around 600,000 billion dollars, where the total global GDP is around 65,000 billion dollars. Theoretically, as the financial bubble bursts, we can use all of our GDP and account for a little over 10 percent of the claims made by the financial industry. One of the core problems with derivatives is that if the bubble bursts, it will prohibit any controlled unwinding of these claims. Instead of the net value, the risk is for the total nominal value of the derivatives.

Moving money around, while each time adding a small (or large) profit margin, might seem a very energy-efficient way to drive economic growth. But it does not create any wealth. As the primary and secondary economies face their limits due to resource scarcity and peak oil, the tertiary economy's trust in infinite growth will eventually start to waver. And without the financial services offered by the tertiary economy, global logistics and international trade will face severe problems. As the confidence level falls, it will be hard to pay for essential consumer goods and energy with virtual financial instruments or imaginative profit assumptions. A crash in the tertiary economy will eventually crash the real economy.

The idea of energy-efficient economic growth is therefore largely an illusion. It is an inverted pyramid standing on the shoulders of the real economy. There have been claims that the economy has decoupled from energy intensity, but these claims call for a more thorough look at what is behind GDP and consumption statistics. Decoupling the economy from energy use will prove to be much more difficult than simply sweeping the risks and costs under the mat with modern financial wizardry and blowing up asset bubbles in the hopes of getting an even bigger fool to pay even more.

The blindness of the marketplace

> Of all the many ways of organizing banking, the worst is the one we have today.
>
> Mervyn King, Bank of England (2010)[4]

If one observes the insane determination with which our financial system moves from one crisis to the next, it is easy to agree with the above quote. But whether we like it or not, we are dependent on the financial sector and the marketplace as they are, for the time being. The mechanisms with which money is created and how it operates also determine what kind of actions and investments our society can make. And who will make them.

According to the signals the markets get, preparing for peak oil is not as profitable as continuing business-as-usual in the short term. Markets make it very hard both politically and economically to invest in the longer term to mitigate peak oil. And even short-term investments rather seek higher returns and larger leverages from derivatives and other financial instruments. This makes it all the more difficult to get financing for projects, even if they were to reduce oil dependency and usage in the short term. Reducing oil use today is usually not profitable, as the benefit one gets from burning oil is higher than the price we pay for it. The infrastructure we have also supports using more and more oil. But when the tide turns, and reducing oil use and dependency becomes profitable on a larger, societal scale, it is already too late. Fixing a machine that is running is nearly impossible.

Trust in the economy

> Ultimately, the economy is, and always has been, a surplus energy equation.
>
> Tim Morgan (2013)[5]

The economy runs and grows largely on trust. The amount of debt, for example, is a direct measurement of how much we trust the economy to grow in the future, allowing us to pay the debt with interest.

Since the tertiary economy, and the real economy with it, grows or stagnates depending on this trust and the willingness to get more into debt, what we are really talking about is a self-fulfilling prophecy one way or the other. If this book has weakened your trust in the economy to grow, you might start saving and paying off your debts. This will cut your consumption in the short term, which will hurt the economy, making the premise of our book come true. But if your trust is in a return to business-as-usual, and in the global economy remaining strong, you can increase your consumption and get more into debt, since you think tomorrow we will have more of everything. And this will make it come true, as your consumption will create more demand, your debt will create more money, and the economy will grow.

But can we simply trust the economy to grow forever? If we keep believing that the economy will grow forever, then will it actually do that? Sadly, no. Unlimited growth in a limited environment, such as Earth, is impossible. Growth that is based on trust alone can be pictured with the example of Wile E. Coyote from the Warner Bros. cartoon. Wile E. Coyote, while chasing the roadrunner, often ran off a cliff. As is quite normal in cartoons, he kept on running even though there was no ground beneath his feet. Everything went really well, until Wile E. Coyote noticed that there was no ground under his feet. What followed was a brief moment as he floated in the air as the realization spread to his brain. When it did, he fell like a rock. His trust disappeared only after he realized that his suspension in the air was not based on anything real.

The real world is not a cartoon, where the animator can forget the laws of physics at will. But our economy does resemble cartoons in the way that it is based on trust, which is not a physical resource and is therefore technically limitless. But the economy exists in the real world. Therefore, it will have to abide by what is happening in the real world, sooner or later. The higher we try to climb on trust alone, the further we will fall, and the worse the damage will be on the way down. One of these falls was the 2008 financial meltdown. The record oil price back then was at least one of the needles that burst the housing bubble. There are dozens of books written on the crash, and even an eye-opening documentary, *Inside Job*, that explains the stupefying irresponsibility of the financial sector in detail. Have we learned from the mistakes made? It does not seem that way. The next asset bubble is being built in the United States on top of shale energy. The first signs that this bubble is bursting can be observed from the asset write-offs that companies have been making in recent years up to billions of dollars.

But the thought of getting less tomorrow remains an unpleasant thought for most of us. This makes opposition to any change easy and natural. And if the structure of the whole system supports our current situation, it makes this opposition even necessary.

Involuntary degrowth?

Involuntary degrowth might be an oxymoron, since the degrowth movement aims for voluntary, planned deconstruction of the growth dependency on our current economic system and society. Degrowth argues that nature and natural resources are limited, and if the system is based on the assumption and aim of limitless growth, it will start to eat itself. If growth is the main aim, as it is today, it will eventually destroy the things that we used to aim for with economic growth, namely the welfare of people and the environment. The aim of our current system is not to grow to some predetermined, manageable size. The aim is to keep growing, forever. If the economy grows by 3.5 percent annually, which is often labeled as *healthy* growth, it will double in roughly 20 years. In the average lifespan of an OECD citizen, around 80 years, the economy will grow sixteenfold. How many more times we can double our economy without destroying our environment, and with it, ourselves?

Degrowth is an interesting concept, but it is also quite unpleasant for the current establishment. This can easily be observed by noting public discussion on degrowth. Degrowth supporters spend maybe 99 percent of their time and effort explaining why the critics are totally missing the point in their criticism, whether through bad rhetoric or a lack of understanding of the concept. Slowly, information has been spreading, making more constructive conversations more common.

When it comes down to it, the question might be framed as follows. Are we willing to sacrifice our well-being, standard of living and environment for economic growth, or are we willing to sacrifice economic growth to maintain at least some of our well-being and environment? The topic is difficult, as historic trends clearly show that economic growth has also increased standards of living. On the other hand, many studies have shown that the actual happiness or well-being of the people has not increased in decades in OECD countries, even though their economies have grown massively.

It remains somewhat naive to believe that the current level of social security and public services could be upheld if we moved to much less consumption. Funding these services is currently dependent on ever-increasing tax incomes, and therefore increasing amounts of energy and materials used. The predicament is that many trends are showing that whether or not to have our current system and culture of increasing consumption is not a choice we get to make. It is a physical reality set by the limits of our planet and environment. What we can choose is perhaps how fast and controlled our move towards a more sustainable future is, and, depending on that choice, how viable that future is for us.

Degrowth is one of those concepts that we need to keep in the discussion. Radical changes often happen only through crisis, and when crisis hits, we usually end up dealing with it with the ideas that happen to be on the table at the time. Involuntary degrowth remains a much harder form of art than degrowth of voluntary, motivated people who understand why they are doing what they are doing.

Collapsing societies

If one tries to bring the possibility of a collapse into public discussion, one soon finds out that it is forbidden. If one wants to discuss collapse, one must discuss the collapse of the Mayan civilization, or that of the Roman Empire. Any analogues to our current industrial society should be kept to a bare minimum. Discussing the possibilities of a collapse remains the domain of bloggers wearing tinfoil hats and other individuals and groups on the margins of mainstream society.

Even in the face of this, and the possibility of accusing this book of doomsday predictions and fear-mongering, we discuss the collapse of societies next.

Mayas, the Romans, and perhaps a few of the other collapsed societies and civilizations we are aware of probably had one thing in common: none of them actively discussed the possibilities and the risks of their own collapse at any point, or thought that it was at all likely. This does not prove that our own society is about to collapse. But it does prove that our unwillingness to discuss the possibility will not stop it from happening either.

Joseph A. Tainter is perhaps one of the more well-known researchers who has managed to bring collapse into the mainstream discussion. According to him, a collapse is not a clear and fast event where the society crumbles to dust in front of people. It is more like involuntary simplification of a complex society. Tainter has observed that this often has to do with the decreasing amount of external energy available to the society. The process of collapse can take centuries, as it did with the Roman Empire.

Jared Diamond is another well-known researcher of collapsing civilizations and ecosystems.[6] He emphasizes humanity's habit of over-using the available renewable and non-renewable resources (this is often called *overshoot*). When the renewable resources can no longer renew themselves, the population will collapse more or less violently. If the dozens of studies on energy, ecology, and the earth's carrying capacity are even roughly correct, we are facing various local crises and collapses quite near in the future. Food, fresh water, biodiversity loss, climate change, population growth, peak oil, and a host of other interconnected problems will spawn local crises around the world at an increasing pace.

What we mean by collapse is in line with Tainter's concept that involuntary simplification of the society happens as the external energy fed into the system decreases. As the energy diminishes, it is simply not possible to maintain a system as complex as before. This fits without much conflict with Diamond's ecosystem views also; as humans use natural resources from their environment, they first and foremost use energy in different forms. Diamond just calls it by a different name. This kind of collapse is hard to notice, especially from within the system.

According to Tainter, humans tend to solve their problems by adding complexity to their society. As one observes, for example, the amount of complexity in the social security system of many western nations, it is easy to agree. The system is so complex that the users cannot use it without support. We have gone so far in adding layers of complexity in our society, that even normal everyday tasks require professional help. These professionals of course require some other professionals in some other area of life. This has led to an increasingly large "life-support" business sector with its own special interests and lobbying machine. Therefore, these largely non-productive sectors of society cannot simply be dismantled even if they were to prove to be counterproductive or useless at some point. Their main job is to make themselves needed.

Simplifying our society

The trends of net energy available to society pictured in this book are quite clear. The amount of oil, and especially the net energy we get from it, will keep diminishing as its price keeps rising. Even if oil is just one of the forms of primary energy we use, it is the most important and most versatile. Following Tainter's train of thought, we might well be facing a reduction in the complexity of our industrial society – whether it is voluntary and planned, or involuntary and unplanned. One of the main tendencies of bureaucracy throughout human history has been to grow and make itself more complex. In addition, simplifying the structure of a system from inside the system is challenging. Seeing the larger picture is difficult, as one "can't see the forest for the trees."

Whether we talk about simplifying social security, health care systems or new legislation, the main focus during the age of peak oil and peak economic growth should be to simplify. It is true that complex structures and systems create jobs, but the usefulness of those jobs can turn negative for society as a whole at some point. From that point on, their main function becomes to make themselves seem as necessary and useful as possible, if for nothing else than the jobs they offer. Most of their actual resources go into keeping themselves busy spinning and working, growing even more and, in some cases, concentrating on making the rest of society's function as hard and inefficient as possible.

Uncontrolled collapse

If we fail to simplify our society in a planned and voluntary way, it will happen by itself when the available energy drops low enough. A mixture of these two is the likeliest event. Greece is a recent example of a fast, uncontrolled collapse. The collapse of the central government and its functions is easily observable. The people's willingness to pay taxes erodes with their trust of those taxes having any useful function. More and more of the economic activity is happening outside any official bookkeeping, in the grey area that does not show in statistics or GDP.

With time, people get used to the new normal where things are no longer done in the old ways. It becomes normal that the level of even routine care depends whether hospitals have resources to stock medical equipment. It becomes normal that only fools pay taxes, as the taxes seem to disappear somewhere without any positive impact. It also becomes normal that the public sector workers get left without a salary and that the people's trust in officials and the central government erodes. A growing part of citizenry takes responsibility for feeding themselves. Local currencies emerge, and different exchange systems replace the old ones. When a citizen starts to grow his own food, societal complexity decreases. The food production industry gets left behind, with its complex logistics and production, and is replaced by more local subsistence with all its merits and drawbacks.

Meanwhile the central government tries in every way to keep the reins in its hands and structures operational, with ever-scarcer resources. Every passing day it starts to concentrate more on justifying its existence. This leads to pay cuts and hardening austerity measures. The citizens respond with unrest, demonstrations, strikes and, if possible, by changing the politicians for other ones. And despite their sugary promises, eventually the new politicians are forced to continue the same cuts and austerity. As the society finally reaches a substantially less complex state, the collapse might slow down and finally stop. This might take decades, so for all practical purposes, constant structural change is the new normal for Greece. Similar change, even if it differs in its strength and appearance depending on the place, might be facing most nations as oil production starts its decline.

Another example from the not-too-distant past is the Soviet Union and its collapse. Dmitry Orlov, who was born in Soviet Union but has lived most of his life in the United States, compares the Soviet collapse to the possible United States collapse in a vivid and entertaining book *Reinventing Collapse*.[7] According to Orlov, the average Russian citizen was far more prepared for the collapse of the central government than the average North American is today. The standard of living in the Soviet Union was much lower, and the society less complex than in modern U.S. Many Russians had been growing part of their food by themselves for years, as their trust in the abilities of the Communist Party to take care of them had crumbled long ago. According to Orlov, the important factor is the level of living standards one falls from. Many poor people in the western world already live a life that would be an utter nightmare for most of the middle class.

The federal armed forces of Germany, the Bundeswehr, released a report[8] on peak oil that tries to perceive the effects that a shrinking economy and instability would bring. Military forces in general of course have an inbuilt tendency to prepare for the worst cases, such as wars, even if the worst never happens. These reports do need to be taken seriously, however, since the military will not spend its limited resources on something that it sees as either extremely unlikely or insignificant in its effects. They also tend to have a pragmatic and neutral view on matters. The report presents the following possible chain of events.

Consumption and production shrink and the economy goes into a recession. Incomes fall even as the prices of necessities increase and production volumes shrink. The ability of the poor to afford food is endangered. National budgets face more pressure as import costs and social security costs rise and the prices of the investments needed to improve energy efficiency increase. Incomes from taxes fall.

The bleak, long-term economic view spreads through society. Companies do not make new investments as they also face a lack of financing from investors, who predict that the companies will not be able to pay interest on their loans or dividends for their shareholders. The banking system, stock exchange and financial markets start to collapse. Banks lose their ability to operate. They cannot pay interest to their depositors as they cannot find suitable investments for those deposits.

Faith in money as a holder of value dissipates, and trust in currencies erodes. The black market and local exchange economies grow. Division of labor, value chains, and processes linked to these, collapse as there is no way to buy or sell unfinished goods in the production chain.

The inter-exchangeability of national currencies collapses, which leads to the collapse of international value chains. Production processes that are not directly linked to consumer goods wither.

Those employed in these chains of production lose their jobs and income. Massive unemployment follows, which leads to further shrinking of incomes for governments. Many nations face bankruptcy. As nations lose access to financial and material resources, critical infrastructures begin to fail. Disruptions in food production and transportation lead to famine.

Rapid systemic collapse

Where the report from the Bundeswehr discussed a slightly slower event, another report from Metis Risk Consulting and FEASTA (The Foundation for the Economics of Sustainability) offers a different view.[9] The report explores the consequences that a crisis in the global financial system will have on the global flow of goods and services. The report acknowledges two main reasons: Excessive indebtedness and unbalances in the global economy, and the end of debt-driven economic growth due to peaking of oil production. The report looks at the global economy through six indispensable subsystems: financing, benefits of mass production, production flows, behavioral models, critical infrastructure and infrastructure for energy and raw materials production and transportation. All of these are deeply interconnected, so failure in one is enough to crash the whole system. Financing is the fastest, and it therefore controls how changes spread throughout the system. Society is left with no chance to prepare. The road to a rapid systemic collapse is pictured as follows.

The eurozone and its banking system face a crisis when Spain, for example, defaults on its national debt. Investors working with massive debt leverage, both in the official banking sector, and in the shadow banking system have to sell their assets in a market situation where there are no buyers. The crisis spreads in the financial system and passes the point where the central banks and governments retain any control over the situation. A self-reinforcing panic deepens and spreads the crisis to a global level. The critical global financial sector starts to waver more and more.

Crashing banks, fears of currency devaluations and defaults, decreasing availability of export loans, and the growing panic crash commercial and industrial flows in countries that are affected most by the crisis. The impossibility of price discovery makes it harder to operate otherwise viable production or implement alternative solutions. Within a week, most factories close their doors, the information network starts to suffer, and the stress and panic in society and government continue to grow.

Soon most companies have to close their operations. The risks on critical infrastructure grow, as those maintaining it cannot manage their own operations and personnel.

Both internal and external trade in the worst-hit countries stop. This causes their trading partners to have problems in their economies. Global trade becomes harder due to the debt crisis. Exports decrease from the most efficient and important commercial hubs. This spreads the crisis to those countries that managed to avoid the immediate problems.

As the commercial crisis passes a critical tipping point, it starts to expand exponentially. In an extremely interconnected economic system it is hard to insulate oneself from the crisis. Eventually the crisis pushes the system that enabled mass production and consumption out of balance. This results in a global systemic collapse of production.

Crisis in trade and its consequences make the financial crisis worse, which passes a critical tipping point in less than 10 days. The economic system and its subsystems lose their long-term balance irreversibly. This triggers systemic collapse.

Local conditions

How and when local events happen depends greatly on the local environment. Societal and political stability, income distribution, homogeneity and demographics of the population, historical factors, economic structures and many other things have an effect on how things will play out. Or if they will play out at all. But in an interconnected world, local events can spread quickly into neighboring countries and areas.

The internet, and with it the ever-evolving social media, and the more and more ubiquitous smartphones with cameras will certainly play their role. But it is hard to say what that role will be. Like any other tool, they can be used to either build or destroy. One thing is quite clear however. The living standard built and enabled by cheap oil will be under great pressure as that cheap oil disappears. It is unlikely that we can go on with business as usual for much longer, even if we avoided the global systemic collapse talked about above. What is the fate of the western standard of living?

Challenges of the welfare society

High living standards and the welfare society are based on spreading national production among citizens, including those who are not very productive for one reason or another. Productivity has grown with automation and machinery. With these, one person can operate vast amounts of power by using external energy sources. With farm machinery, one farmer gets the strength of thousands of people. This has led to the current situation where a small percentage of people grow and harvest the food for everybody. The other people are free to specialize as doctors or nurses, engineers, teachers, salesmen, stock brokers, and so forth. In an automated factory, one employee can oversee and manage massive production facilities and quantities.

The actual welfare that people are experiencing is dwindling, and it has less and less to do with financing the whole social security system and its services. By concentrating on growth alone, its negative side effects tend to also grow. These include stress, mental health problems, drug abuse, the deterioration of social capital in society, and financialization of everything.

Productivity is dependent on external energy

Growth in productivity is mainly due to mechanization, automation, and feeding external energy into the processes. Of this energy, more than 80 percent is non-renewable fossil fuels. Without this energy, nothing resembling our current society would have been possible. What keeps us believing that anything resembling our society will be possible after fossil fuels?

The mechanisms are the very same ones that have served us so well this far:

- belief in the technological process, belief that there is always enough time, and in the endless ingenuity of the human race;
- belief that the past trends will keep on going in the future;
- belief that someone else will solve the problems for us, or that they will solve themselves;
- belief that the threats are not real.

There is no denying that humans have been, and are, ingenious. Technology and ingenuity are not energy, however, and an innovation by itself does not produce anything. On the contrary, technology usually uses energy. We have been very ingenious in finding more and more ways to use more and more energy as fast as possible. Predicting the future from the rear-view mirror on the other hand is not very smart, as it only works when the road is straight. Technological development will also no doubt have a big role in the future, but it is not something that we should blindly trust. People saying that the only way for humanity is to grow and develop is just that: people saying. They are in no way laws of nature.

What makes the predicament hard to observe and even notice, is a lack of knowledge among the general public of the role and importance that energy plays in our society and everyday life. We lack energy literacy, so to speak. This makes it easy to offer people solutions based on nothing else than make-believe. It also makes it difficult for people to realize the scale of the problem, let alone challenge the business-as-usual thinking. For example those that want to replace fossil fuels or nuclear with renewables are often fond of saying that "we have an almost infinite amount of renewable energy available." As one measures the amount of sunlight reaching the earth, one can conclude that this is indeed true: there is a vast amount of energy shining on us all the time. The problem is that what we need is energy in a dense, practical form. It is called *exergy*, or the ability to do useful work, which sunlight is not. Similarly, there is a large quantity of water in the morning grass, but it is hard to quench one's thirst with it.

It is essential to ponder what will happen to the modern welfare society, as the energy making it possible becomes expensive and scarce. For most European countries, most of their fossil fuels are imports. If they stop importing oil, their society will grind to a halt and crash, so that is not really an option. With their dependency on imported energy, they are also dependent on a successful export sector.

The rising prices and potential limits on availability of oil and energy are not the only problems in the coming years. Further challenges are, depending on one's location, presented by:

- demographics, as the large post-Second World War generation reaches retirement;

- the growing demand for health care by the above generation;
- the trends in stress-related problems due to overworking, which is leading more and more people to drop out from the workforce before retirement age;
- a growing amount of debt both in absolute terms and as a percentage of GDP, and the growth in the costs of managing that debt as the interest eventually starts to rise;
- the problems in the global economy that will make exporting goods more challenging;
- as growth slows down, the things dependent on growth facing problems, such as real estate values in certain areas, the expected returns on investments for pension funds, and so forth;
- the effects of progressing climate change, such as the increasing uncertainty of crop yields and damage due to increased amount of storms.

The list of problems we are facing is long, even without peak oil and all its risks and effects. Of course one can see problems also as possibilities. From a local point of view we can have a big impact on our future if we make good choices today. As we move forward, oil efficiency and dependence become ever-larger factors for a society and its local viability.

The welfare society has to streamline itself and the services it provides. Simplifying social security and cutting counterproductive bureaucracy are two obvious solutions. Now, more than in many decades past, it would be important to work together for a common goal. This can only happen if we function as a single group, not as many factions with each having their own special interests. Having an ever-growing gap between the poor and the rich is a recipe for worsening societal unrest. The CEOs cannot raise their salaries by tens or even hundreds of percent while at the same time demanding pay cuts for their workforce (or refraining from pay rises) for the good of the company and its competitive advantage.

The situation in Europe seems gloomy, as the financial crisis has likely only been a warm up. Many experts see two options: either the European Union falls apart in one way or another, or it gets tighter and moves towards a Federation. The trust and solidarity of the participants will be challenged, as for example the northern countries' food production subsidies, that were meant as temporary fixes, will be questioned. Meanwhile there will be growing support for domestic food production.

Would a tighter union be a better way to mitigate for peak oil? There are valid arguments both ways, and a lot depends on how and how fast the process would go either way. EU politics are often seen as inefficient, bureaucratic and slow. This does not add to the attractiveness of deeper cooperation and a tighter Union. Many nations will have an increasing interest in holding on to their strategic assets and economic sectors. On the other hand, disintegration will increase the risks for both domestic and international conflicts. Extreme groups and views have historically found their source of power in general discontent.

Politicians and the shrinking pie

What kind of decisions would politicians make today, if their premise was a future of very slow growth? And what will happen to debt payments and societies if the assumed growth, always lurking in the near future of economic forecasts, fails to materialize?

The basic justification for more government debt has been to avoid overly hard cuts that would imperil the seeds of future growth. This is a valid justification, although it is debatable (as is the assumed wisdom of the marketplace) how efficient government and the public sector are at using that debt in a productive way. It is often slow and hard to change policies that are firmly set in place. Society can also rapidly change how it operates. This is called revolution, and it is something many nations might face if their governments do not react fast enough, or turn out to be inefficient.

But what if the current trends continue? What if the global economy fails to grow as expected, but oil and energy costs keep rising? Europe goes from one crisis to the next, all the while losing its ability and resources to act decisively. It would be preferable if the debt taken today were used to invest in benefitting society in the longer term. Debt-driven consumption really means consuming today from tomorrow. As the future is unfolding, this seems more than a little irresponsible. But the temptation of just a little more debt is strong, as it means no cuts or austerity just yet. But the resource and energy situation keeps getting worse.

The situation is hard for anyone, but it is especially hard for a politician, whose aim is to be re-elected every few years. Where will cost cuts come from, if the shrinking tax base demands it? Which cuts will not endanger future growth, private consumption or investments made by companies? Which cuts or tax increases are OK for the politician's key supporters and political backers? How will they explain to the people that tomorrow they might have to consume less than today? And can these changes be done in a way that will not scare people and encourage them to start saving too much, which would wreck the consumption-based economy, at least in the short term? It would be hard to avoid losing one's popularity. In this thankless situation, rare political leaders (and citizens) are called for, who see the common good as more important than their personal welfare or career.

If we want to mitigate the future problems caused by peak oil, it would be sensible to direct long-term investments into projects that lessen society's dependence on oil, and make its use of oil more efficient. One option to consider is to investigate new infrastructure and investments through their assumed oil efficiency. Does a project make the local or national oil-efficiency or oil-independency better or worse? At the same time, the population should be informed more openly, even if there is a risk that they start to take action to improve their situation.

Who would divide the smaller pie in the best way?

Dividing a growing pie of wealth is much easier than dealing out austerity. As long as there is even slightly more for everyone, even a growing income gap does not feel so bad. But who should be doing the dividing of the wealth so that society as a whole benefits the most? The markets? The bureaucrats and civil servants? Politicians? Special interest groups and lobbyists?

The marketplace puts its trust in price discovery. Whoever has the ability to pay gets the fuel. From society's point of view, this is not an optimal way to go about things. The total damage to society is much greater in a situation where the rich continue sailing around in their yachts as the poor have to stop working because they cannot afford to buy gasoline to drive to work. In this case, the poor cease to be productive members of society, and move to live on social security or on some other means. Inequality and social exclusion grows, which will eventually lead to increasing social unrest and decreasing national security.

In a more sudden crisis, as when a state of emergency is declared for example, the absolute power moves to current politicians, officials and the national defense forces. These kinds of situations are becoming more and more likely, and recovery from them harder and harder, with the peak of oil production and other natural resources. In Europe, it is already becoming the new norm that there is some kind of crisis going on all the time. This saps the time and resources that society has available, leaving other problems unsolved and even unnoticed. The central point is that we need to think and plan ahead, not when the situation is already upon us. The results with planning ahead are an order of magnitude more preferable to results from hasty improvisation in an emergency.

The peaking of oil production is however only one of the symptoms of our current way of life. The current way that modern society functions includes short-sighted aiming for personal gain, glorifying economic growth (and sacrificing everything else for it), an insane level of hyper-consumerism that our marketing industry helps to create (which also creates economic growth), and our almost complete inability to see the consequences of our actions even one generation ahead. The system has been crippled by the system itself. Multinational contracts, organizations and corporations are so powerful that changing the legislation or otherwise going against their interests is difficult, if not impossible.

We have also built many skillful pyramid schemes around us. We are, or at least we appear to be, dependent on keeping them growing forever. For example, retirement funds alone demand perpetual growth to stay operational. In addition, environmental protection usually (not always) becomes important only when other needs and wants have been met. It is hard to voluntarily give up benefits once they have been acquired. It is even harder if everyone else is not doing the same. Continuing to grow forever, on the other hand, is impossible.

What cannot go on forever will eventually stop.

The officials and peak oil

> All truth passes through three stages. First, it is ridiculed. Second, it is violently opposed. Third, it is accepted as being self-evident.
>
> Arthur Schopenhauer

> When the critical mass understands the true state of things, the shock can be extremely violent – especially if it is magnified by an oil shock or similar event. Whole industries can face rapid extinction.
>
> Philosopher Tere Vadén

We have followed the international public debate and media coverage on peak oil for years. Some notes have been presented elsewhere in the book, but perhaps it is good to draw together some central conclusions here. There has been little mention of the Export Land Model (ELM) in any official documents. The reality still remains that the peak in net exported available oil to the open markets was reached in 2005. Experts, official reports, and the media have not mentioned it much, aside from the researchers themselves who came up with the model. Many media reports still confuse the notion of peak oil, where the subject is production volume, with the amount of total oil reserves. Not many mention the quality and composition of the remaining reserves – it is simply assumed that bitumen or kerogen can be produced as easily as conventional oil, as long as the price is right. Price will not overcome physical barriers, at least not in the short term. This is leading us to a supply crunch, probably during this decade. The worsening EROEI (energy returned on energy invested), along with the growing risks to short-term supply disruptions that can lead to crisis, have not been mentioned that often either.

The other thing that is not fully realized is the timing of peak oil and the magnitude of its effects. It is somewhat controversial that many scientifically literate people seem not at all worried about peak oil, as the oil-producing companies and exporting nations simply tell them there is nothing to worry about. There has been peer-reviewed literature on the subject, and the picture one gets from there is somewhat different. Many, if not most, environmentalists tend to downplay peak oil and its urgency. There always seems to be enough time to prepare, and the deeper consequences seem to remain hidden.

It is anything but irrelevant when the actual peak is and when and how we start to prepare for it. This is one of the key points we have tried to present in this book. Some central observations:

1 We will have to adapt, either voluntarily or involuntarily.
2 The more we postpone any planning for mitigation, the more painful that adaptation will be.
3 Voluntary, planned mitigation will take decades for transportation alone.
4 Mitigation will require oil, as most of our tools and machinery that we use to do something about the problem, run on oil or are made of oil.

The public debate about peak oil, when there is one, is still largely on a level where:

- People argue what the term actually means and remind that there have been warnings about it for decades.
- People belittle the effects and risks associated with peak oil.
- People point to assumed massive resources in oil sands, oil shale and shale oil.
- People argue that the market and technology will take care of the expensive oil problem.

There is also a quote from a Saudi Arabian oilman that one sees from time to time: "The Stone Age did not end for the lack of stones. And the Oil Age will not end for the lack of oil." This cheap rhetorical trick hides the fact that actually we have plenty of examples in our history where a local civilization, population or ecosystem has collapsed due to overconsumption of energy and resources. It is also true that we are currently using more stone than ever before, including the Stone Age.

Considering the meager growth of around half a percent during the last few years, one can assume that peak oil volumes might not be radically larger than the production volumes of today. This means that we are in reality living in peak oil right now. Producing unconventional oil economically is the result of high oil prices, in addition to improving technology. This places an effective floor on oil prices, making it hard for the prices to fall significantly from what they are today.

But this is the case mainly in the longer term, if the economy continues to function relatively normally. There are many mechanisms that might, in the coming years, cause the price of oil to fall significantly from current levels. Most of these are tied to falling real incomes (purchasing power), especially when considering discretionary spending. High oil and food prices, combined with stagnant or falling real wages, make mass consumption possible only by increasing debt. Debt can only be increased up to a point, that is, until managing that debt becomes too costly. Discretionary spending is hit with a triple shot of falling real wages, higher oil and food prices and higher costs of managing debt. This leads to the fourth hit, where people in discretionary industries start to lose their jobs, making all of the above mechanisms even stronger. This way of lowering oil prices through collapsing economies will likely wreak much more havoc than high oil prices ever would.[10]

At the community or city level there seems to be much less knowledge about peak oil. Zoning, which is usually under local governance, is usually made on the assumption of cars and more cars. This practice has been established through decades of practice and industry lobbying, so it is hard to even see any alternatives. In many cities any other practices have even been made illegal or impossible with regulation, for example, by regulating a certain amount of parking spaces per household. The prices, often tens of thousands of euros or dollars, of these parking spaces are often paid for by the households whether they have cars or not.

For the average citizen, it is the local level that offers opportunities to make changes on a meaningful scale. As part of the background study for this book, we made an official inquiry through some council members to city officials in Lahti, Finland. In our inquiry we asked how the city was prepared for rising oil prices and what kind of costs the city and surrounding area had due to their oil use (both direct and indirect). Even though the answers we got were shallow and did not really answer the real questions, our inquiry had other effects. Some city officials realized there was a problem they had not been aware of, and they are now actively researching the situation and trying to find ways to mitigate for it.

The local grassroots level is also the place where many projects are born. A great example is the multitude of local resilience projects that come under the umbrella of the Transition Movement. Working locally means that real results can be achieved with relatively meager resources. The scale of operations is ideal in many senses, as the local environment is usually where people spend most of their lives. It is small enough so that an individual can have an effect, but large enough that the effects actually count toward something. Transition Towns usually increase the viability of the local neighborhoods or towns in the case of a worsening situation due to peak oil, climate change and economic downturns. There is a lot of literature on the Transition Movement where one can find ideas on what to do and how.

The invisibility of peak oil in the media and public debate

If peak oil is a significant problem, then why has it not been discussed much in the public media? This could be due to the following things:

1 Information and news are usually supportive of the views of a smaller, if more powerful, part of society.
2 News media are usually corporations, for whom it is more profitable to offer topics that sell well or interest their advertisers.
3 There is little unbiased and reliable information available, and, even where it is found, it is unusually difficult to simplify to an easily readable form due to its complexity.

The IEA has been criticized for making overly optimistic predictions on future oil production. There is also evidence, as presented in the story below, that this optimism might not have been accidental. Lionell Badal, a student at Exeter University in Britain, unfolded a chain of events when he got a confession from an anonymous IEA official in 2009.[11 12]

Back in 1998, the IEA found out, to its surprise, that a peak in oil production would be reached before 2020, and possibly as early as 2014. The group working on the annual *World Energy Outlook* report could not publish this finding straight away. It ended up coding the message into the report as *unidentified unconventional oil*. It would appear from out of nowhere around 2020 to increase the total production. There were only a handful of people who understood the message.

As the oil predictions from the Energy Information Administration (EIA, part of the U.S. Department of Energy) seemed to be in conflict with those of the 1998 IEA *World Energy Outlook*, the IEA was pressured to dispel the group that had worked on the publication. The IEA got back in line and knowingly continued to publish misleading reports under political pressure. In 2004 it predicted that the price of oil would gently edge towards 29 dollars per barrel by 2030. According to anonymous IEA sources, many people inside the organization believed that it would prove to be impossible to get production above 90–95 mbpd. It was political pressure from the United States that was behind these distortions. The United States wanted to prevent widespread panic in the financial markets, and wanted to maintain its control of the overall oil situation.

Fatih Birol, the current Chief Economist of the IEA, was the only member of the original 1998 group that was allowed to stay in the organization. Recently Birol has been publicly vocal on the impending oil supply crunch. Even if the IEA denies that any of this conflict existed, many others such as *The Guardian*[13 14] have confirmed the chain of events. According to Lionel Badal, he has also given talks on the subject to the EU Commission, but has anything actually changed?

Nearly everyone in the western world has heard of the greenhouse effect. It is no wonder, since it was first observed in 1824 by Joseph Fourier. The Intergovernmental Panel on Climate Change (IPCC) was founded back in 1988, over 25 years ago. Its mission is to gather scientific knowledge and estimate the effects of anthropogenic climate change. Over 650 experts and researchers took part in IPCC's 2007 report. The first United Nations Framework Convention on Climate Change was signed in 1992. In spite of all this, our carbon dioxide emissions have grown by 50 percent since then. The meaning of climate change for the majority is still attached to things such as how much snow there was last winter or the personal, often economic, inconvenience of doing anything about climate change.

The peaking and scarcity of oil is, compared with climate change, researched very little by largely scattered researchers and small groups in some universities. Related topics, such as geology, are of course researched well and taught widely. Maybe the most significant organization is ASPO International. It is a network of national ASPO chapters and does not get much official support from any government as far as we know. So if our fight on climate change has, despite all the resources poured into it, remained quite toothless, how can one expect the awareness of peak oil to spread?

What makes researching peak oil relatively hard is the lack of reliable information. Many nations hold their reserve estimates secret and unconfirmed by third parties. It is no wonder that the estimates on the remaining reserves vary so much.

Peak oil in politics

It is quite possible for politicians to create, or at least appear to be creating, green business opportunities around climate change mitigation. Green growth, green-collar jobs and clean tech startups sound nice and create the appearance that a lot is being done about the problem. Oil scarcity on the other hand tends to limit economic growth, and preparing for it would mean extra costs for almost everyone. These topics have not been easy or popular as election themes even during good economic times, let alone in the prolonged economic downturn currently underway. The truly difficult subjects do not tend to get much publicity and most voters remain ignorant of them. There is no demand for discussing more complex topics from the public, which makes the election of aware, smart, and skillful political leaders less likely. The unexpected events, the so-called Black Swans, are not recognized, which makes preparing for them impossible.

Nassim Taleb, the author of *The Black Swan: The Impact of the Highly Improbable* (2008),[15] notes that there will always be different black swans, but by becoming aware of them, they can be turned grey. He is also skeptical because of the unforgiving role of the messenger. If we assume that a leader predicts and steers his country away from an event of small probability but big negative impacts through large national sacrifices, then he will probably be harshly criticized for overreacting and reckless spending. Why? As the negative event never occurred, it is hard to know what its impacts would have been, and even if the threat was real at all.

At the risk of generalizing, it can be assumed that the bigger the problem, the less talk of it there will be, and the more likely it will come as a total surprise. But why do people not even want to hear about immediate threats like this?

There have been studies, for example in Alberta University,[16] on what makes people interested in the world around them. The results have not been very encouraging, as there is reduced interest when:

- the subject is more complex (for example the effects of oil on the economy, climate, food prices, human behavior and biodiversity loss);

- the subject's implications are more widespread (for example oil in geopolitics);
- the event will hit sooner (for example climate change in 2050 is easier than peak oil in 2015);
- they know less about the subject beforehand.

So the more information there is on peak oil and its ramifications, the more likely a person is to surf to easier subjects and entertainment. When researching the effects of communications about climate change, it has been noted that the message will stay vague and scattered if the political parties are not consistent with each other when addressing the subject.[17] This, of course, is one of the main drivers behind climate denialists' success in spreading doubt. This also applies to peak oil, although there has been consensus in many countries to consistently not discuss this topic at all.

In the business world, the event horizon usually remains at a few months, or at most a couple of years. In our experience, the understanding of peak oil and its societal effects among business leaders and CEOs is often quite narrow. It goes as deep as the effects of more expensive gasoline, and leaves it there. The economic contraction that peak oil is causing is hard to understand. Only a few seem to recognize the possibilities and business opportunities. Research studies or high level reports seem to carry little weight if they are in conflict with the established business-as-usual views.

We can conclude with this somewhat surprising finding: When especially scientifically literate people get more information on climate change, they tend to fall ever deeper into denial if they have a previous, conflicting view on the matter.[18] People tend to have an uncanny ability to cherry-pick only the information that reinforces their previous beliefs (known as "confirmation bias"). This raises the question of how meaningful writing this book has actually been . . .

Glimpses at the cloudy future

This chapter offers some scenarios on the future that peak oil is taking us towards. They are not predictions of any sort, but rather some narrow views on how things might pan out, looking at the following:

1 Oil efficiency of transportation
2 Inside the crisis
3 The time of polarization
4 The fall of consumerism
5 Climate change after peak oil.

The setting can be any industrial country. The details and chain of events are determined somewhat by local conditions. The point is to give some tools on how to think about and perceive the situation both globally and locally.

Oil efficiency of transportation

As the volatile oil prices continued their inexorable climb and the local crises started to pop up against everyone's wishes, the government started making radical readjustments. Sensible actions were taken only after wasting a lot of precious time and resources on different alternatives and options. The people were showing ever clearer signs of dissatisfaction everywhere, and it was just a matter of time as to when the first truly dangerous demonstrations would begin.

One of the first steps was the loud and clear endorsement of ride-sharing and public transportation with both carrots and sticks:

- Tax deductions and reimbursements for using one's own car were cut to the minimum.
- Congestion payments were implemented in large cities and parking fees rose sharply.
- Road lanes that had congestion problems were reserved for buses, taxis and ride-sharers.
- The media started to show ads and news on smartphone apps that could be used to find and book ride-sharing.
- Employers were forced to continually reduce the fuel use of commuters.

Car dealers were forced to report the total costs of car ownership with the average annual driving distances. This helped to factor the otherwise hidden fuel costs into the purchasing decision of a new or used car. Many people living in cities or suburbs, who still had jobs and money, bought small and light electric cars and scooters for their daily needs. The multitude, however, settled for repairing their aging vehicles and bought a bicycle.

In the background, preparations for rationing gasoline were started. The opposition was loud, but it was somewhat mitigated by making the "Fuelcard" have no actual limit at first, nor did it affect the prices. It was merely a preparation so that rationing would be more effective should the need arise. The next phase was to implement a progressive fuel tax with the Fuelcard that ensured some amount of cheap gasoline or diesel for everyone. By making the progression steeper, it was possible to control total fuel use, and cars with small engines and good mileage increased in popularity. The system also saw a feature that enabled willing people to sell their share of low-tax fuel to the highest bidder. As such, fuel rationing proved to be an effective way to control total consumption. It also offered a form of social security for people who could do without personal cars, as they could sell their quotas.

As the trend in passenger cars and driving turned south, plans for new highways were quickly scrapped. The resources were directed to maintaining the current network of roads and expanding the railway. Speed limits were lowered permanently to save fuel.

Jobs, services, and settlements were brought closer together and nearer to existing transportation networks by new zoning. Small, local stores started to make their inevitable comeback. Faster internet connections for the existing, more rural settlements started to appear, as more and more essential services went online. The faster networks also found new users with the increasing amount of telecommuters.

Biorefineries proved to be economically problematic. When the price of oil had risen high enough to make them economically viable on a commercial scale, it was already too late. The government helped to kick-start some projects, but the political wringing and special interests behind the projects took much of the already precious time and resources.

Regulation was loosened on public transportation. As a result, many bus companies started using various forms of pre-booking, dynamic pricing, and flexible departure times to increase the amount of passengers and reduce prices, especially compared with personal cars. Hybrids and natural gas powered buses started to appear as fast as the companies replaced and increased their fleets.

The rural gas stations, which had the necessary infrastructure and bus stops nearby, started to promote cheap all-day parking with electric chargers for electric vehicles, a weekly or monthly car washing service and other packages for commuters to help with their otherwise diminishing business. This enabled the commuters from the suburbs and exurbs to travel the short distance to the local gas station and continue with public transportation from there.

The trucking companies faced the double pressure of increasing diesel prices and diminishing volumes of goods being transported. Goods were moved to railways as much as possible, and even rivers and lakes started to see more local freight. Postal services dropped to 1–2 times per week in many remote locations. Local and community-powered courier services appeared to fill the gaps. The newspaper and magazine businesses started to offer bundled free electronic reading devices for those who did not order the paper version.

The political cost of mitigating and adapting to the rapidly changing environment was high. Many long-term solutions were hopelessly late and proved to be too much for the debt-burdened state and private sector. The immediate benefits were seen as too small compared with their costs. As awareness of the impending scarcity of oil and other resources spread, many projects did not find supporters. There simply were not enough resources.

Many special interest groups, labor unions, and political parties demanded that fuel taxes should be lowered to make gasoline and diesel oil more affordable. Tinkering or removing the progressive fuel taxation system became a permanent topic in elections. But the system as a whole offered a working solution to mitigate shorter crises and to control the total fuel consumption of the passenger traffic. People were simply forced to prioritize their driving, but the changes could be implemented in smaller steps. This helped public transportation to adapt to increasing passenger volumes.

Inside the crisis

Peak oil, the credit crisis, and massive indebtedness finally forced governments to cut their costs dramatically. As the global economic crisis was prolonged, cheap international credit disappeared. The trade balance was already strongly negative, and the volatile price of oil kept climbing. The resulting structural change started pushing working tax-payers from the private sector to the receiving end of the social security system. This created an ever-worsening situation for public sector workers, as their wages were paid ultimately with taxes collected on private sector production. Many found themselves in an awkward situation, where the safe government job had disappeared and left them with a skill-set that was not very relevant in the marketplace.

As the crisis progressed, a work obligation was declared for those receiving social security. Many found themselves kind of employed, but without a salary, doing more or less light manual labor in workshops, building sites or helping the overburdened public services sector. "Do your share" became the new government slogan, and, to everyone's slight surprise, many did.

Social services and health care faced steep cuts. Patients were moved away from hospitals to their homes, where basic care was often provided by quickly-trained, otherwise jobless people working for their social security. The professional health care staff were assigned to go around the neighborhood, taking care of patients needing more advanced health care. Social security moved to handing out coupons that one could use to get potatoes, flour and other non-perishable food from distribution centers.

The maintained traffic network got tighter, and new construction was done to modify its condition. Energy-saving renovations and transformation of office space into residential buildings got cheap labor from the unemployed. Manufacturing was strictly zoned to be beside existing railways or harbors, but also closer to residential areas. Smoke chimneys seen from the kitchen window started to signify jobs close enough to cycle to, instead of signifying unpleasant smells and pollution.

The maintenance of roads was dropped to the bare minimum, and the resources were directed to improving railways, light rail, and other energy-efficient infrastructure. Those who wanted to live in a remote rural home had to pay the full expenses themselves if they did not work in farming or forestry.

Maintaining parks and public buildings was all but halted; aside from fixing life-threatening places or maintaining the central city areas for tourists. Suburban, small plantations were given to people living in nearby cities; not so much to ensure adequate food availability, but to activate people and to help build and strengthen local communities. Community colleges started to offer more courses on permaculture, urban gardening, and the like.

The state also started to manufacture the "citizen bicycle" that was relatively cheap and durable. They also made an electric bicycle that was equipped with a small trailer. Many jobless people found themselves earning added benefits from hard labor working as couriers with the trailer-cycles.

The auto-industry saw some tightening of tax progression based on fuel efficiency, but otherwise it was largely left alone. Electric vehicles were temporarily tax-free, their popularity grew, and car sizes got smaller. In addition to bicycles, many small and light four-wheeled vehicles took to the roads, which did not allow high speeds anyway due to the declining condition of roadways.

Distance learning, after faltering for a while, became popular as the fiber optic networks spread and wireless capacity grew. The children of farmers and such were offered a lighter and more practical education if they wanted to continue with their parents' business. Later this was applied to many other trades. The brightest and most talented kids were given special education for free. In time, higher education became the prerogative of the wealthy elite, as the middle class kept shrinking.

Crime was on the rise due to social restlessness, declining living standards, and increasing drug abuse. Public opinion forced prisoners' conditions to be lowered to below that of the regular jobless working for their social security. Prisoners also found themselves working longer days and doing work based more on their skills. Immigration laws were made tighter as a result of political pressure.

The time of polarization

The general chaos brought about by the oil crisis was harsh for those who thought they had saved enough to retire comfortably and even leave some inheritance. Many trusted the system and official propaganda just long enough so their savings and wealth lost most of their value.

As a result, people were largely put in the same situation where the living standard was chiefly the result of everyone's ability to muddle through the ever-changing situation. The youth were the exception. They had not had time to buy their own home and get into debt before the credit market along with the housing market collapsed. The youth still faced growing unemployment though.

Another exception was those who had moved their wealth to safer places. The state protected land owners as long as that land was producing efficiently. Those with large amounts of visible wealth in stocks, funds, or real estate found out that they were easy prey for increasing equity taxes.

A few had prepared beforehand by laboriously making their homes and lives as energy-efficient as possible. This offered them relative comfort, along with the sometimes spiteful envy of their neighbors. A bigger problem was the fifth of the population that refused to acknowledge what had happened, and tried to stick to their lifestyles tooth and nail. For most, this was only possible at the expense of others. As a result, many types of crime increased: embezzlement, theft, violence, vandalism and economic crimes that people used to vent their despair, frustration, and bad luck on their surrounding environment.

Some neighborhoods tightened their communities to get shelter from the ever more erratic society. This could easily be observed from the local state of things; as public waste management and other services disappeared, those neighborhoods lacking any communal sensibility soon started resembling slums with all their ill effects. The tighter communities took care of their environment, and even organized local security patrols in addition to cleaning and maintenance.

In the rural areas especially, the importance of community grew, as those in the countryside had always had to rely more on each other anyway. This was made easier by the fact that farmers and land owners got to keep most of their wealth. They also had better access to liquid fuels, which sparked both envy and cooperation. Almost all the communities turned more or less inwards, which sometimes led to xenophobia and other problems.

Labor unions found themselves in a situation where they would either try to use their power or let themselves drift in the chaotic situation, going extinct sooner or later. The tensions between employers and employees rose ever higher. Strikes and difficult demands on the political system made it difficult to make the trade balance less negative. The blame for everything was, as always, directed at someone else. Populist rhetoric became more and more common.

Immigration and births started to be observed from a totally new, austere perspective: new citizens were new mouths to feed instead of being valuable human resources who would add to production and well-being.

The fall of consumerism

After a decade or two had passed from peak oil, and the bumpy plateau of oil production was a thing of the past, the worst chaos in society started to abate. Resources in general continued to become scarcer and local crises happened all the time, but his had all become part of the new norm. People remembered the old, insane days, when everything was consumed without any thought of moderation with mixed feelings. Back then, even talking about rationing had been a bad joke. The youth blamed their parents for their self-indulgent air travel around the globe for their annual or even monthly vacations. The older generations remembered the old days with thoughts of longing and guilt. For many it was tough to see their children serving rich Russian and Chinese tourists.

Europe continued to import oil from Russia, the Middle East, and north Africa. The export revenue of most countries lagged behind import expenses of oil and other resources. This had dropped oil imports to less than half of their 2013 levels. A network of bio-refineries was built in Europe that used all available feedstock and waste streams to their last crumb, but still there were liquid biofuels only for very limited uses. Electricity and district heat was made largely in nuclear power plants, since natural gas and coal were expensive imports in many countries, and there were other uses for natural gas in manufacturing fertilizers and as fuel for transportation. More and more firewood was used in houses, even if its price kept rising. Protests against nuclear power became less common as the price of electricity kept rising. With the economy and financial sector largely broken, it was hard to get funding for energy investments.

Previously, the value of food, energy, and other critical goods imports had been around half of a nation's export value. In the new situation, this share rose to nearly 80 percent in many countries. Factories and businesses drove their machinery beyond repair, and the citizens tried to fix and mend their appliances as best they could. Home appliance stores went bankrupt, as they were built on the assumption of fast turnover and planned obsolescence.

Oil imports were increasingly directed straight to critical uses such as farming, forestry, mining, construction, industrial production, public transportation, and transporting goods. They were the only sectors where oil use did not drop very much from 2010 levels. But this meant that there was very little left for anything else. Oil used for personal transportation dropped by 90 percent, and oil burners in houses and buildings were changed to other technology at record pace. Heating oil as a product largely disappeared. The roads started to see vehicles with a variety of power sources from electric to natural gas to even wood burning, but cars as such got much scarcer. Traffic accidents all but disappeared, and bicycles became tremendously popular, even in winter time.

Climate change after peak oil

Humanity managed to wriggle through peak oil for a decade or two longer with all possible substitutes. Oil sands, natural gas, shale oil, and deepwater oil along with liquefied coal managed to offset most of the decline in conventional oil production, but the price was high. The promises of growing production the industry gave and the critical role of oil in societies led to political paralysis. This led to insufficient mitigation and ever-loosening environmental regulation on pollution and carbon emissions. Nobody dared to disrupt the permanently fragile economy. And if some brave politician or CEO dared to try, he was quickly replaced with another. The developing countries bought an increasing share of the available oil, while the industrialized countries fell ever more deeply into recession, debt crisis, unemployment, rising living costs with falling income, and eventually into social unrest.

One of the biggest losers was climate change mitigation. National and international leaders had their hands full with more immediate problems, and climate change grew into an ever bigger elephant in humanity's living room. Even talking about it started to become a sign of bad judgment, lack of broader understanding and even as anti-humanity. Center stage in the public debate was given to energy security and the rising civil unrest caused by the widening gap in living standards. Local collapses in society inevitably led to local collapses in ecosystems, as people raided what was left of the natural resources.

The emissions budget that researchers had calculated to give us a reasonable chance to stay under 2 degrees of warming was overrun without much notice. Even if humans failed to reach the speed of burning fossil fuels that IPCC scenarios had predicted, it was more than made up by other actions humans took. The sixth great extinction shifted into high gear as the remaining arable land was converted to monocrops for biofuel production. Rainforests and boreal forests were cut down and burned for electricity or liquefied. Increased fishing, along with seawater acidification, saw the rest of the fisheries collapse along with many local marine ecosystems.

Well before mid-century, positive feedback loops related to climate change started to kick in. Permafrost in northern peatlands started to melt, releasing massive amounts of methane and carbon into the atmosphere. There was a huge campaign to dig ditches to dry the peat in a vain attempt to stop the methane from being released. In the end, people ended up digging up the peat and burning it, which only made matters worse. The permanent ice cover melted away in the polar regions, causing more and more heat to be trapped within the planet's darker waters. As the oceans continued acidifying, their ability to bind up atmospheric carbon dioxide decreased. The Arctic was free of summer ice for the first time in 2020, and in a decade it became the new norm. The race to exploit Arctic oil and natural gas deposits intensified. The oceans continued to rise at an increasing pace. During the second half of the century, there was a race to develop methane hydrates, not so much to get the methane, as to stop it from being released into the atmosphere. It threatened to kick the already wildly uncontrolled climate change into a totally new gear.

The net energy that humanity still scraped from fossil fuels was spent mainly trying to fix the results of different catastrophes due to new, unpredictable weather patterns; panicky adaptation to the new planet was underway. It was not the same planet that the elderly, now feeling both guilty and desperate, were born into. The carbon-free society planned just a few decades ago had turned into a bitter fight for the remaining resources in a hostile environment. The high standard of living that had been widely used to rationalize the burning of fossil fuels a few decades ago was but a vanishing dot in humanity's rear-view mirror.

What should one do?

Various examples, ideas and even policy suggestions can be found throughout this book. Depending on one's personal situation, the feasibility of some actions can vary greatly. What works for one, does not necessarily work for another. What works today, does not necessarily work tomorrow. Keeping this in mind, we refrain from giving straight "do this and don't do that" kinds of suggestions. Instead, we will try to give a framework and tools to think about and prepare one's personal situation for the coming trends. After these, some thoughts on the future of employee or entrepreneur markets are offered.

Life style

A citizen living in the developed world is in many ways dependent on the seamless working of the society around him or her. Thinking about and mapping out these dependencies can help in finding ways to reduce them. What are the basic needs you have, who supplies them and how? What are the risks attached to them? How could you manage without a car? Without vacations and flying? Without a regular job? All of these are, of course, everyday realities for many already.

Moving to a cabin with a potato farm, cabinets full of ammunition, and canned food is not a very good solution either, at least for most humans. Living alone in some cabin and staying totally dependent on the surrounding society leaves a vast area of pragmatic thinking, practical analysis, and increasing one's resilience in between.

Think about your job. How essential is it now and in the future? The most important thing is to be good at what you do, but of course some sectors will be more vulnerable than others. It may be wise to learn extra skills now, while there is still a chance to practice and develop expertise. When a crisis hits, there will be no time for learning or preparation. Many will feel paralyzed, and most of the energy and time will go towards dealing with the crisis. Even if the crisis is not sudden, the constant structural changes happening around us will demand time and resources.

Try to improve your ability to deal with the change. Resilience, or a certain kind of flexibility, in the face of a changing environment is very important. When your own situation has improved, try to observe your immediate surroundings: your workplace, neighborhood, family and relatives, hobbies and other people who have an effect on your life. A good life requires a good living environment.

Political realism

Things in society tend to move forward based on short-term gain and the needs of different special interest groups. The space for rational long-term planning has been shrinking. Also, seeing a world where one's own position is threatened is hard. Accepting truths that challenge one's own position and income in society is nearly impossible for many, and usually it is harder the more one has to lose. Not many want to even hear about peak oil and its dynamics. The rhetoric tricks and biased data and reports used time and again to avoid any meaningful discussion is proof in itself.

Economical point of view

You need to plan your economic activities with a slightly longer time frame in mind. The markets and general environment are still largely unaware of or unwilling to see the coming trends, and therefore continue to give distorted signals. You need to think for yourself, and think a bit further down the road than tomorrow. Trusting the social security network is not a good strategy, and neither is blaming this or that politician. Getting around the official system in different ways can be justified. The system as it is in many countries might cease to function or it might lose the justification for its existence. Local trading and cooperation, neighbor help and the gray economy are all different aspects and degrees of operation outside the formal economy. It remains for the individual to evaluate what is justified and what is not. For many people, this evaluation is based on the fairness and integrity of society as a whole.

Where should you invest any savings? Should you take on more debt or raise your level of savings? Should you invest now, or wait for a better chance? Can you modify the energy costs for yourself or your company to be more flexible so that you can cut down on your expenses easily and without a radical drop in your living standards if incomes fall?

The affordability of oil, energy, and other essential goods can be very different from the trends the market is seeing. The less this comes as a surprise, the better. Lifestyles made possible by cheap petroleum will no longer be possible for most, with scarce and expensive oil. It will take a while to get used to this idea but, after you do, you will have a much better starting point for planning what is essential and what is not.

Lifestyles can be altered significantly without much drop in actual living standards. But this can only be done beforehand and with careful planning.

Job market of the future

> But what if the world we've been living in for the past five years is the new normal?
>
> Professor Paul Krugman (NY Times, 2013)[19]

What sort of workers will find jobs in a future where:

- The price of oil, and with it energy-intensive transportation such as trucking and flying, will become less affordable.

- Agriculture, food production and logistics chains may need to scale back on their dependency on imported energy.
- The incomes of the middle class will not grow, and the share available for discretionary spending decreases.
- The relative price of food and energy keeps rising.
- Many European nations face increasing problems as their population gets older, societies get more divided, and crime and social unrest increase.
- Economic growth as we have come to know it will not return even in the longer term.
- Information moves rapidly over long distances, but people and goods move slower and in smaller quantities.
- The geopolitical center of gravity will move from the west eastwards and southwards to Russia, the Middle East, China, Brazil, and others.
- Globalization slows to a crawl and civil unrest increases.
- The probability of local crises, environmental catastrophes and supply disruptions of oil increases.

The forthcoming structural changes in society and business will take away jobs and livelihoods from an innumerable amount of people. Their old vocations will simply not be affordable any longer. Of course this is not a new situation, as whole sectors of the economy have faced extinction in human history, and there have been crushing economic slumps in many places. But now the situation is more global than ever before. Now there may not be a torrent of cheap energy and resources to get us back onto the growth track after a few years of restructuring and creative destruction.

How the changes will progress locally depends greatly on the local system and actions taken. Problems caused by excessive debt in Greece are causing a systemic collapse there. Schools are being closed because of an inability to buy heating oil. People are importing firewood from Bulgaria because they cannot afford oil to heat their homes.[20] As most European countries continue along their current path of growing inequality, it is all but inevitable that social unrest will grow. Countries with more equal incomes will do better and stay more stable, which will in turn attract investments and businesses looking for less risky places to do business. As stable societies become less common, those that remain stable will get a competitive advantage.

Worrying about one's job, whether one sits at a desk for days on end or toils at ground level, is justified. And if one's career has not been chosen yet, it would be frustrating to go to school for years only to find that there is no job market. The following thoughts and points should be read bearing in mind your own strengths and possibilities to secure a living in the future. Every profession will have employment available for the best, but not everyone can be the best. Therefore, it might be smart to avoid professions likely to suffer the most.

Primary production

As oil becomes less affordable, raw materials such as metals become more expensive to mine, refine, produce and transport. This will make recycling more competitive. There is a lot of work to be done to enable effective recycling for all those materials that currently are not recycled and reused properly. Research and development of recyclability of products should become more important.[21] Renewable materials such as wood, and products made from it, should become more popular.

Local ways to produce and transport food might prove to be more cost-effective than the current practices. Finding ways to reduce the oil and energy use of the food sector as a whole, preferably without cripplingly high investment costs, will be valuable. Gardening and growing food in cities in various ways will become more popular.

Local production and refining of energy, such as biogas, will become more popular especially if it can be used to replace imported oil.

Energy and infrastructure

Wood and other biomass will be collected and transported for energy. This is most effective when transportation distances are quite short, which will employ construction industries, manufacturers and maintenance. With rising energy costs there will be orders for energy renovations, counselling and appliances that help save energy. These will, however, face the reduced ability to get loans for bigger investments.

Local micro-production of energy is also likely to grow in popularity. Even if solar panels or some other energy solution are made somewhere else, there will be the need for sales, installation and maintenance. All of this will also require massive new infrastructures to be built and realignment of how our society is built. If this is done, the resources will be taken away from somewhere else: new roads, shopping malls at remote intersections, or building more public sector offices and buildings are some likely options to see reduced demand.

Manufacturing

Manual labor has been disappearing for roughly two decades. It has disappeared to China and automatization. As we are forced to realize that we simply cannot move production somewhere else and keep consuming as if nothing happened and that the rising costs of transportation and debt management will make this impossible sooner rather than later, this trend will shift. Starting up and retrofitting old (and new) facilities will demand skilled engineers, mechanics and maintenance personnel.

There is a whole world of designing and manufacturing products that concentrates on material and energy-efficiency during their operational lifetime, not forgetting easy reparability and upgradeability. This kind of research and development requires a high level of understanding of the product from a multitude of viewpoints and sciences. It is time to put our high level of education to its proper use.

Home and school

The trends of second homes in the countryside, excessive renovations, and updates to interior decoration will hardly continue at their present pace. This will have negative effects in many sectors from construction, design and retailing, to advertising. The exception is with renovations that increase energy-efficiency and investments on land in different forms from forestland to small vegetable gardens. Consumer electronics might also be a somewhat surprising exception, as staying at home playing or watching movies is a reasonably energy-efficient way to spend free time compared with many alternatives. From society's point of view it is also a good way to offer the "circus" to people and keep them in their homes if they have lost their jobs and previous hobbies.

The public transportation networks, along with bicycle routes, will increase their effect on everyday life in the local neighborhoods. The trend of individualism might start to reverse, as communities need to find support for themselves, instead of from the state or some other official means. Local sharing networks and timebanks will grow even if they do not offer much monetary income to people. Taxing this kind of work and sharing is both difficult and unpopular, but the increase in social capital can be huge, if hard to measure.

Schools may find it useful to increase distant learning courses at most levels. This will require adaptation from all the participants, from hardware to software, and from students to lecturers. This will not replace the networking that happens especially in higher education, but it would supplement the current lectures as more people would be able to participate.

Transportation

So far, the trend in physical transportation has been to use more energy to move faster. As energy becomes more expensive, this trend might reverse. This will slow down globalization and make the world a bigger place once again. Personal driving will decrease, which will lead to an aging fleet. This will be good news for car repair and maintenance shops, but bad news for auto manufacturers and dealers. Retrofitting a gasoline engine with ethanol or natural gas conversion kits may grow in popularity where these fuel options are available.

Jet fuel will make up a bigger share of the total costs of flying, and economic hardship will make flying for both pleasure and work scarcer. Weekend shopping trips to New York or annual vacations to distant beaches in Asia will become a thing of the past for most of the middle class. Commercial aviation will face severe contraction that will leave the nimblest, most efficient and less indebted companies ruling the skies, but with much less volume. Businesses that depend on growing or current levels of commercial aviation will also face trouble as volumes shrink. These include airports and their personnel, and those getting their livelihood from mass tourism.

But what will grow? In personal transportation it is easy to see an increase in, and diversified uses for, cycling. As people give up personal passenger cars, different services such as taxis may be able to avoid radical collapse even if they see a cut in their business clientele. They will also have to compete with increasing ride-sharing and car-pooling services. With GPS, the internet and smartphones, these options will become much more practical than they were during the last oil crises in the 1970s.

Rising fuel prices will see to it that long-distance trucking and flying will lose some of their competitive edge in terms of freight. Rail and water transportation will increase as fast as the infrastructure allows. Logistics as a whole is dependent on increasing consumption volumes, so it will shrink as a sector. It will also be forced to be more and more efficient in a more and more unstable business environment. Shortening production chains will also decrease the need for logistics, while more local storage capacity will be needed.

Services

The services that stand on the shoulders of a growing and wealthy middle class will be under pressure. As the middle class sees their real income shrinking and the prices of essential goods rising, they will not be able to consume as much as before. This will mean a longer circulation time of nearly all durables from clothes to home electronics to furniture. This will mean problems for manufacturing (usually done somewhere else) and sales of these items.

Building and maintaining the information network's capacity requires energy, but it can also help save energy. Services that reduce the need for moving around will especially have a competitive advantage. Services and content that can be consumed through the internet will also have more and more competitive edge, having relatively lower production prices than physical goods or media. As internet services grow, the energy efficiency of server centers becomes a bigger issue. People will tend to move from hobbies that use a lot of energy to hobbies that can be done and enjoyed at home. From a writer's point of view it would be desirable if some of this enjoyment was gained through reading. Reading has seen its growth shift towards electronic books, especially in the United States. Europe will most likely follow, but the market is more fragmented.

Tourism

Mass tourism is as energy-intensive as it gets. Basically the idea is to take as many people as fast as possible to a location as far as possible. There they will spend as much money and local natural resources as possible in as short a time as possible. Mass tourism is also dependent on a wealthy and large middle class.

Long-distance vacations are likely to see their long-term growth trends shifting, so it might be a nice idea to have that trip somewhere far away and really enjoy it while it is still affordable (if the emissions caused have not been a problem thus far). Optionally it could be smart to invest one's money in a more energy-efficient lifestyle. People will still have the need to get away from home, at least for a while, so local and domestic vacations will have the possibility of enjoying growth.

There will of course be wealthy people flying between Asia and Europe in the future. St. Petersburg has five million inhabitants and businesses of all income levels and investment capital willing to bring some of the oil and gas money back. However, the current and past trends of growth need be looked at with a critical eye when planning to build that new spa or skiing resort.

Russia's influence and interaction in Europe is likely to grow. This will mean that the Russian language, culture, and customs will increase in value in the job market, and not just in tourism. The huge injection of wealth that Russia has enjoyed for the last decade or so has already been seen in tourism from Russia. This will likely continue for some time.

Public sector

European people are generally getting older, which will mean problems even without peak oil. The circumstances will be such that an increasingly smaller part of the population will have to provide for a growing part of the population, while the energy that has so far enabled productivity and growth will get scarcer and more expensive. Economic growth will also, most likely, largely disappear. Eventually the public sector will have to adapt by shrinking, as the motivation of the private sector to fund public services will decrease. This will, and already has, proved to be unpleasant, with large corporations already slipping from tax payments into tax havens. As the realization spreads that the young will most likely never enjoy meaningful retirement benefits, so their unwillingness to pay for the benefits of current retirees with grow. Further, as dividing the shrinking pie actualizes, the buying power of current benefits will most likely be eroded. The welfare services offered currently might shift towards self-service, maybe with as much automated social and health services as possible.

On the other hand, the growing and voting elderly population will aim to have their interests represented in politics. This could mean pressure to increase various services offered to retired and old people. As the financial basis for those services is already crumbling, there will be tension between generations.

Security

As social unrest grows, the risk of conflicts of different scales also grows. Those who have the resources (and reason) to increase their personal security will be tempted to do so. The same applies on a national level, although the battle for diminishing resources will make it harder to increase national security and armaments. Increasing nationalistic rhetoric from the political sphere will help rationalize this. The widening income gap, joblessness, shrinking public resources for social security, and therefore a growing need for people to escape reality with drugs and alcohol will create demand for both public and private security services. Those individuals who do not have too many moral or ethical constraints will probably find employment in international security.

Finance and savings

The financial and banking industries, along with the real estate industry, have enjoyed the good life during the booms in recent history. This might change as debt reaches levels where it can no longer grow and people start to worry about the stability of derivatives, funds, and other investments. This can lead to increased savings and investment in more crisis-proof places and objects. The problem is that safe and profitable investments tend to be scarce. At the end of the fossil-fueled growth trend the financial sector might re-evaluate its need for labor. On the other hand, bankruptcies and economic hardships will employ bookkeepers, debt counselors, lawyers, and debt collection services.

What we have done about this?

We have, obviously, written this book. It was written because it needed to be written. It was written so that as many people as possible could better understand the current situation, our dependency on oil, and the many risks and consequences it will have and is already having. The original version that this international adaptation is based on, *Finland after Oil*, got a lot of attention. It got nominated for both of the top Finnish non-fiction book awards. First, it was nominated for the national *Kanava* non-fiction book award among ten nominees. After that, it was also nominated for *Tieto Finlandia*, the top Finnish non-fiction book award, among six nominees.

We have tried to mitigate our lives and careers to be less oil-dependent and more financially independent. But of course we still enjoy the many benefits that oil makes possible. We have given speeches, lectures, and hosted conversations on the topic for people at the local pub, university students, and politicians in the chambers of power. We have given statements to political committees. We have written numerous articles, opinion pieces, and blog posts to educate the public for magazines and newspapers of most, if not all, political persuasions. We have given interviews for papers, radio stations, and television. We have probably burned a lot of bridges while criticizing the decision makers, but hopefully we have built some new ones as well. We have also founded Peak Oil Finland, a non-profit organization that has been a kind of umbrella for many of the things mentioned above (at www.peakoil.fi). Peak Oil Finland is applying for membership of ASPO International.

This book is not a premise for us to sell our own solution, or a marketing tool for a start-up. This does not mean that such a start-up does not exist – one has to do something to solve at least part of the problem, and even make a living. Harri Paloheimo and Heikki Waris worked at Nokia in Finland for many years, and there tried to get Nokia to start developing mobile solutions to help mitigate peak oil. Since that proved to be futile, another route was selected. The most prominent and immediate problem is road transportation. It uses more than half of all the oil, and yet we are as a society utterly dependent on it. Private road transportation is the most ineffective, and also the most vulnerable, so it was a natural choice. The solution needed to have:

1 Viability. Sponsors and grants will disappear when push comes to shove.
2 Scalability for large crowds. A feel-good solution for green hipsters was not enough.
3 Feasability. Users cannot be expected to radically change their habits or lives.
4 A real impact, preferably on the double-digit percentages on oil efficiency.
5 Fast availability, without large investments, and with the devices and tools that people currently have.

Ride-sharing services meet all of these criteria, and there are already companies that enable use of the internet, text messaging, and smartphones to make ride-sharing easier. The next step, still in development, might be a dependable and easy-to-use form of ride-sharing for goods. Almost 20 percent of driving is done to and from shopping, and the potential for savings is even greater than ride-sharing for people. It is hard for two families to share a ride to a store, but one car has enough space to transport goods for many families. By helping one neighbor, the oil use for that trip is halved, and helping two neighbors cuts it to one-third. It is also preferable to transport stuff rather than people. Stuff does not share its moans about where the world is heading, nor does it ask you to change the radio station you are listening to.

The developed ride-sharing platform allows people to easily and safely organize transportation where it is needed. From the store to the home, from the post office to the home, from home to a recycling station, and of course between friends and internet auction sites for example. All the participants save in various ways – not just the driver or the recipient or the store, but the whole economy and the environment. The bigger ride-sharing gets, the more is saved. Those interested in participating can register at www.piggypaggy.com. It is a small step for the individual, but it will help in bringing one more solution to the inevitable to your area.

Afterword

If we are to see and solve a problem, we first need to become aware of it. Even if peak oil is not directly the reason for many problems, it is indirectly behind most of them. The world of expensive and scarce oil is also running into a glass ceiling of economic growth. On the other hand, with excessive indebtedness, demographics, the widening gap of living standards, and so forth, the glass ceiling will be there eventually anyway – most likely sooner rather than later. Many of the problems feed on each other, and also make their interconnectedness hard to see.

During the writing of the original book (2011 and 2012) and this international edition (2013), the situation has mostly become worse. Climate change has progressed unforgivingly with extreme weather becoming more common and the Arctic ice caps melting to record lows. Climate negotiations on the other hand have gone nowhere.

It has also been interesting to see that some conclusions drawn in the book have gone from the fringe to the mouths of mainstream economists in just a few years. An increasing number of politicians, officials, and analysts have said that the European economy is in for a longer period of very slow growth. In 2013, this talk has gone more global, as even China has been downgrading its growth forecasts time and again. The solid trust in "healthy" growth that the politicians had as recently as 2011 has been turning into a scramble for this and that monetary policy that should return us to the path of growth. Most still fail to see the real problem and therefore look for the wrong solutions. The problem is that energy and especially the energy returned on energy invested (EROEI) that actually runs the economy, is decreasing. This is clearest with oil, which is the most important source of energy and the fuel of choice for the global logistical system.

The shale oil and gas boom in the United States started to make global waves with the IEA *World Energy Outlook* of 2012. The Europeans have started to scramble for their supposed shale gas deposits, but so far the results have been dismal. Amid all the hype, people tend to forget about the reality. Shale gas cannot replace oil as such. Shale oil is not cheap oil. Shale production is so young that it is hard to see how the production will develop, but the IEA 2013 World Energy Outlook put its peak in the United States to be likely during this decade. The same conclusion had already been drawn by reports based on actual production numbers instead of corporate marketing material.

The growth forecast of global liquids production has been small for several years. Even this small growth requires that everything goes extremely well, from Iraq's almost unprecedented growth in oil production, to Russia fixing its infrastructure and economy to meet western standards, to shale oil not being a bubble to robust economic growth materializing from somewhere. According to recent reports and white papers, oil prices need to double in 10 years if we dream of continuing even a meager 1 percent growth in oil production. It is doubtful that the economy can afford such prices, as it is already failing to afford even the current ones in many places.

With all the economic worry, climate change has become less popular. The truth, based on physics instead of policies and negotiations, is that we would actually need to leave roughly 80 percent of the known fossil fuels in the ground. In real terms this means that burning coal would need to stop today. The fact that it is practically and politically impossible does not affect the physics that rule the climate one bit. Physics does not negotiate, no matter how hard we would want it to. Humanity is well on its way to several degrees of global warming with an increasing risk of positive feedback loops kicking in, with a system where it becomes less and less possible to undertake any meaningful mitigation or adaptation.

As long as the central aim of a society is to grow the economy by any means necessary, humanity will keep banging its head against a brick wall. This means that the volume of material consumption will swell, which will demand ever-growing use of energy and other natural resources. A future worth living requires that we drastically increase the efficiencies of energy and material use, and the utilization rates of infrastructure. When we change the rationale for our actions – for example, when we aim for a decent living standard and well-being for as many as possible with a minimal ecological footprint – then we can begin to find solutions.

Notes

1 PIIGS is a term often used by financial analysts and news media for the countries of Portugal, Ireland, Italy, Greece and Spain. Given the somewhat derogatory nature, its use has been discouraged, by the *Financial Times*, for example.

2 Stieglitz, J. (2012), *The Price of Inequality: How Today's Divided Society Endangers Our Future*, W.W. Norton & Company.

3 Morgan, T. (2013), *Life After Growth: How the Global Economy Really Works – and Why 200 Years of Growth Are Over*, Harriman House.

4 www.economist.com/node/17363435

5 Morgan, T. (2013), *Life After Growth: How the Global Economy Really Works – and Why 200 Years of Growth Are Over*, Harriman House.

6 Diamond, J. (2004), *Collapse: How Societies Choose to Fail or Succeed*, Viking Adult.
7 Orlov, D. (2011), *Reinventing Collapse: The Soviet Experience and American Prospects*, New Society Publishers.
8 Available, for example, here: www.energybulletin.net/sites/default/files/Peak%20Oil_Study%20EN.pdf.
9 Tipping point: near-term systemic implications of a peak in global oil production. Report available at: www.feasta.org/2010/03/15/tipping-point-near-term-systemic-implications-of-a-peak-in-global-oil-production-an-outline-review.
10 http://ourfiniteworld.com/2013/11/15/whats-ahead-lower-oil-prices-despite-higher-extraction-cost.
11 www.theecologist.org/News/news_analysis/437079/how_a_22yearold_student_uncovered_peak_oil_fraud.html.
12 http://permaculturenews.org/2011/05/12/how-the-global-oil-watchdog-failed-its-mission.
13 www.guardian.co.uk/environment/2009/nov/09/peak-oil-international-energy-agency.
14 www.guardian.co.uk/business/2010/aug/22/peak-oil-department-energy-climate-change.
15 Taleb, N. (2008), *The Black Swan: The Impact of the Highly Improbable*, Penguin Books Ltd (UK).
16 http://sc.teachers.ab.ca/SiteCollectionDocuments/ASEJ%20Vol%2042%20No%201.pdf, page 16.
17 www.scientificamerican.com/article.cfm?id=popular-opinion-on-climat&page=2.
18 www.nature.com/nclimate/journal/vaop/ncurrent/full/nclimate1547.html.
19 http://topics.nytimes.com/top/opinion/editorialsandoped/oped/columnists/paulkrugman/index.html.
20 www.bloomberg.com/news/2012-12-18/greeks-can-t-find-euros-to-buy-heating-oil-with-winter-economy.html.
21 More can be read on this subject in *Cradle to Cradle: Remaking the Way We Make Things* (2010) and *The Upcycle: Beyond Sustainability—Designing for Abundance* (2013) by William McDonough and Michael Braungart.

Appendices

Appendix 1 – Oil in numbers

It is hard to follow the oil news and keep things in perspective if one is unaware of the underlying facts, including real numbers on the state of oil resources and oil reserves. These numbers below do not aim to be absolutely accurate since they have been rounded to be easier to remember. They are in the ballpark, and that is what is often important. Some numbers, like prices, also fluctuate yearly so that needs to be kept in mind.

Units and terms

- 1 liter of oil contains around 10 kWh of energy, which corresponds to 2 weeks of human labor.
- Speaking on a global scale, the normal unit is mbpd, or million barrels per day.
- A barrel is the common unit of oil. One barrel contains around 159 liters of oil, or 42 US gallons.
- 1 ton of oil equals a little over 7 barrels (7.1–7.4 barrels, depending on the quality and density).
- 1 barrel contains around 1.7 MWh of energy (1700 kWh or 6.1 gigajoules).
- 1 barrel has the energy content of around 17,000 hours of human labor.
- Per volume, oil has about double the energy content of coal, and 4 to 5 times the energy of firewood.
- A resource is totally different from a reserve. A resource means the total amount of oil in place (original oil in place, or OOIP). A reserve means the estimated amount that can be produced with current prices and technology.
- 1P (or P90) oil reserves have a 90 percent probability to be produced at current prices and with current technology (1P means *proven*).
- 2P reserves (or P50) probability is 50 percent (2P = *proven* + *probable*).
- 3P reserves (or P10) probability is 10 percent (3P = *proven* + *probable* + *possible*).

- Usually 20–60 percent of the oil in a conventional oilfield can be extracted. This is called the *recovery factor*.

Oil use in Europe and elsewhere

- France uses around 1.69 mbpd
  - This is around 4 liters per person per day
- Germany uses around 2.36 mbpd
  - This is around 4.7 liters per person per day
- The UK uses around 1.47 mbpd
  - This is around 3.7 liters per person per day
- The Netherlands uses around 0.93 mbpd
  - This is around 8.8 liters per person per day
- Italy uses around 1.35 mbpd
  - This is around 3.5 liters per person per day
- Spain uses around 1.28 mbpd
  - This is around 4.3 liters per person per day
- Portugal uses around 0.23 mbpd
  - This is around 3.4 liters per person per day
- Greece uses around 3.1 mbpd
  - This is around 4.6 liters per person per day
- Russia uses around 3.17 mbpd
  - This is around 3.5 liters per person per day
- China uses around 10.22 mbpd
  - This is around 1.2 liters per person per day
- The U.S. uses around 18.6 mbpd
  - This is around 9.3 liters per person per day
- Canada uses around 2.4 mbpd
  - This is around 11.5 liters per person per day
- Brazil uses around 2.8 mbpd
  - This is around 2.3 liters per person per day
- Japan uses around 4.71 mbpd
  - This is around 5.9 liters per person per day
- India uses around 3.65 mbpd
  - This is around 0.5 liters per person per day
- Saudi Arabia uses around 2.94 mbpd
  - This is around 17.2 liters per person per day.

All oil consumption data is taken from BP's Statistical Review of World Energy 2013. Population data taken from Wikipedia.
The global oil use in 2012 was distributed to different sectors as follows:

- 53.4 percent – Transportation
- 13.6 percent – Petrochemicals
- 5.8 percent – Other industry

- 8.7 percent – Buildings
- 6.3 percent – Power generation
- 12.1 percent – Other (agriculture, bitumen, lubricants and other non-energy use).

Oil globally

- Humans use around 90 million barrels of oil products every day.
- This amounts to over 30 billion barrels or 4 billion tons of oil annually; over 1,000 barrels are used each second.
- Annually, around 10.6 billion tons of CO_2 are released as a result of our oil use (2009).
- The U.S. uses almost a quarter of this at slightly below 20 mbpd (over 10 liters per person per day). Two-thirds of this is used in transportation.
- The EU uses around 13 mbpd.
- The average person in the world uses around 2 liters per day.
- More than half of global oil is burned in transportation, around 1 liter/person/day.
- Around 95 percent of transportation is powered with liquid fuels, usually refined from oil.
- Russia, the U.S., and Saudi Arabia are the three biggest oil producers, each producing over 10 mbpd.
- The production from current oilfields depletes around 4 mbpd each year according to the IEA.
- It means that we have to find and bring into production (or develop old fields to keep them from depleting) a comparable amount of new oilfields within Saudi Arabia, Russia or the U.S. every 2–3 years, if we are to maintain current levels of production.
- If consumption is to increase 2 percent annually, we would need an additional 2 mbpd of new production every year.
- With an annual increase of 2 percent, the current estimated peak production of around 100 mbpd would be reached in 5 years.
- During the last 8 years, the average increase of all liquid fuels has been less than 1 percent, marking a shift in the previous trend of 2 percent annual growth.
- According to BP (2013) the total proved reserves in the world are around 1650 billion barrels. Of these:
  - around 400 billion barrels are extra heavy oil and bitumen in Canadian oil sands and Venezuela's Orinoco Belt;
  - around 300 billion barrels are so-called *political reserves* in OPEC countries that have not been proven by any outsider.
- According to UGES (Uppsala Global Energy Systems), the available reserves for production during this century are around 1,500 billion barrels. At the current speed of production this is enough for around 50 years.

Appendix 2 – Oil markets

The global marketplace is believed to be an efficient and rational system, where different actors can participate in a more or less equal way. The reality is somewhat different. Some central terms and their backgrounds are presented below.

The market is about exchange that is done at agreed prices and volumes. The amount of information that affects these two factors is enormous. This information includes business news reporting, information regarding the development of different market instruments that have to do with oil, statistical reviews and predictions on production, consumption, and oil deliveries, and various press releases of different actors in the marketplace.

The market functions optimally only if most of the actors have similar levels of information. Certain computer algorithms bring significant advantage to some actors, as they can process public market data much faster than any human. This in itself distorts the market's optimal function. In addition, certain actors hold closely much of the vital market information. This includes estimates on national and international oil companies' oil reserves and ability to produce oil, and various countries' strategies to limit oil demand or add pressure on certain producers to produce more or less oil. In addition, oil producers can make oil delivery contracts outside of the open market, which results in less oil ending up on the open market. This makes it possible for a well-informed large exporter (currently, for example, Saudi Arabia or Russia) to sell its production at the futures market when prices are high (as some of the predicted production fails to enter the market) and then buy it back when prices fall (as some other actors have given up).

Some of the mechanisms, information sources, and tradable items are described below.

Production, demand, storage levels

Future prices in the trade of raw materials are estimated on the basis of variations in demand, production and current levels of stocks. As demand is bigger than production, storage levels fall, and price increases are to be expected. For price discovery in oil, the levels of consumption and production (*flow*) play a bigger part than levels of storage and reserve capacity (*stock*). For example OPEC can create expectations for lower or higher levels of production in the near future to keep prices within a certain range. Optimistic estimates can be used to increase hype on certain oil substitutes' future production. Especially at the state level, actors can steer expectations to the desired direction by emphasizing or downplaying the risks related to civil unrest or threat of war.

Spot and futures

Oil is traded at the daily *spot*-price. Many users of oil want to plan their activities with a longer term perspective, at least a few months into the future. This is what the futures-trade is for. They are essentially contracts to buy or sell oil in the near future. The normal situation is called *contango*. It means that oil delivered in the future is slightly more expensive than oil at the spot-price. This is normally due to costs in storage, risks of future deliveries, and assumed increases in demand. If the future prices are lower than the spot prices, this is called *backwardation*. If the one buying or selling oil is not an actual consumer or producer of oil, these futures need to be *rolled-over* before their due date to a date further in the future. Roll-over means that some actors are forced to buy or sell certain oil futures before a certain date.

Delivery standards

The delivery contracts are usually made on certain standard *oil-markers*, which serve as price guides for buyers and sellers, of which the two most well-known are Brent and WTI. Brent was originally based on certain North Sea oil produced by Shell, and nowadays it is produced at around 100,000 bpd. The spot-price is a price for Brent delivered on a certain date, and announced by the oil reporting company Platts. It is a single price used directly for 60 percent of oil deliveries, and indirectly it affects an even bigger share of oil trade. It should be noted that in 2009 Saudi Arabia, Kuwait and Iraq started using a price derived from the Argus index (ASCI) that is calculated from three different Gulf of Mexico oil categories. WTI, which was used before, was distorted by local events and did not represent the global situation.

Indirect oil-wealth

Indirectly, there is much more wealth that is based on oil. This wealth includes the stocks of oil companies, IOUs, and their derivatives such as *credit default swaps* (CDSs). The estimated secure cash flow of the oil companies through these massive amounts of nominal wealth is maintained in investors' books. The continued revenue and profit from these papers is essential for their holders and their existence, and through them, to the functionality of the whole financial sector. As oil is produced from riskier places, risk increases over the profitability of oil companies, and therefore for system-wide write-offs. For example in the accident at the Deepwater Horizon, the U.S. government set an upper limit for the liability of BP to contain the economic side of the crisis.

Oil exchanges

The trade in oil futures is made in exchanges specializing in raw materials and commodities, such as ICE and NYMEX. It is also possible to speculate on oil prices through other exchanges or diversify one's investments by investing in ETFs that deal in commodities. Most participants use third parties as brokers. There has been speculation that these third parties use their insider information to make trades on their own behalf (costing their clients).

Trading outside the exchanges

In addition to public trading, many investment banks broker oil trades and delivery contracts on behalf of their customers. This is called *over the counter* trading (OTC). It is not regulated, and therefore only legislation on private contracts applies to it. Trade can also be done outside the public commodity exchanges. The largest brokers have superior levels of knowledge on both public trading and OTC trading. They can also work as advisors in investment banks, and be otherwise involved in the oil business. This makes it possible for them to collect money from ETFs that assume (based on public data) that they will collect hefty profits from their futures only to find out that the assumed amount of oil at assumed prices is not available on the markets at the set date, because part of the oil has been sold outside of public trading.

Payments and transfers

Oil is traded in U.S. dollars. One exception was the embargo on Iran. Even trading between China and Russia has been made in dollars. The actual payments are made as international transfers, using the SWIFT system.

The markets are always wrong

Sometimes one hears the saying that the markets are always right. This can, with as solid reasoning, be turned on its head to say that the markets are always wrong. If the markets discounted events and presumed trends into the prices, they would develop in a calm manner. A small news item would not cause big volatility in the prices. Many say that the markets are more an embodiment of group psychology. This is also true, but it needs to be remembered that markets are used to redistribute vast amounts of nominal wealth. As such, markets are a zero-sum game. The many factors listed above offer some actors many opportunities for manipulation and leverage.

It is also worth saying that currently the oil futures do not see very far into the future. They are thus largely unable to discount the availability of oil reserves or our ability to substitute oil in its many uses when a serious scarcity occurs. The result is that oil is currently likely to be sold at a deep discount from the point of view of future generations. This fundamental fact allows anyone who grasps it and is able to move their current consumption to the future (meaning able to invest in oil efficiency), will gain significant economic advantage in the long run.

Appendix 3 – Reason alone won't make change happen

Why is it so hard for humans to think rationally?
The vast majority of people react more often on emotion, even though they would like to justify their actions later with rationality. Psychology and behavioral sciences know more than a few mechanisms that affect the thinking and acting of us all. By becoming more aware of them, their negative effects can be mitigated:

- Resistance to change and its psychological features (anger, denial, resistance) are one reason why an outside change such as peak oil causes an opposing reaction with many people. It takes time and active effort to overcome this.
- *Confirmation bias* means that humans tend to look mainly for information that confirms their previous opinions on any matter. If we have been told for decades by authorities that oil supply will not run out, we have a pretty strong bias to think that this must be so. This will not make resistance to change any easier to handle.
- Belief in authorities is related to the previous bias which enables us to externalize difficult matters and decisions to somebody else. We also tend to believe, wrongly, that this "somebody else" will make decisions that somehow favor us.
- *Cognitive dissonance* means that we tend to ignore the consequences of our actions that can create conflict between short-term enjoyment and long-term inconvenience. It is obvious that every drop of oil burned today would have more value in the future.
- The *tragedy of the commons* is a theory linked to cognitive dissonance. According to this theory, the benefits for an individual outweigh the common good. This is because the benefit is collected by the individual, whereas the price is paid by everyone together. It makes it easy for us to reason that it is not worth reducing one's driving, as someone else will burn that oil anyway with nothing left for the future. The fact that we know people are selfish stops us from behaving unselfishly. We assume that someone else will be selfish on our behalf.
- *Natural fallacy* means our inability to distinguish between how things really are and how we wish they were. One example is the common view that human ingenuity and technology will solve all our problems, no matter whether the problems are the result of the economic system,

overpopulation, or the scarcity of fossil energy and resources. The fact that this has often actually happened does not make it a law of nature.

- *Fallacy of composition* means that it is hard for people to see the relationship of a single event to the whole. If one stands up in a football match to see better, does it mean that if everybody stood up they would all see better? Or if you are able to buy as much gasoline as fits into your car, does it mean that the same is true for everybody, everywhere?
- *Groupthink* means that it is hard for us to derive opinions (and actions) from the common views of our peer group, even if we sense that they are in conflict with the information we have. If your relatives, neighbors, or friends have built their lives around driving and cars, it means that cars and driving are common subjects in discussions. A life without a car would cause social friction as it would deviate from the norm. The group would expect you to show that your new lifestyle would not make your company uncomfortable.

Index

1970s oil crises 52–56

agriculture 152–161
- environmental impact 154–155
- Europe 156–157
- fertilizers 154, 155
- genetically modified organisms 158–159
- green revolution 4–5, 153–155
- issues 159–161
- organic production 158
- problems 155–156
- United States 157–158

Alberta, Canada 36, 80–82
Aleklett, Kjell 137, 152
Arab Spring (2011) 35
Arctic oil production 78–79, 138
asset bubbles 198–199
Association for the Study of Peak Oil and Gas (ASPO International) 22, 214
austerity 209
Australia 22
aviation 148–150

Bakken formation, North Dakota 84–85
banking sector 156, 204, 230
batteries, electric vehicles 119–120
Berman, Arthur 110
biofuels 99–105
- ecosystem risk 103–104
- global production capacity 101–102
- growing of biomass 100–101
- problems 102–104
- production 99–100, 102
- social implications 103

biogas 112–113
biomass to liquid (BtL) 93
Birol, Fatih 12, 213
BITRE *see* Bureau of Infrastructure, Transport and Regional Economics
bitumen production 79–83
Blackout (Heinberg) 97
The Black Swan (Taleb) 24, 26
boron cars 148
BP (British Petroleum) 16–18
Brazil 77, 236
BRICS countries 169
BtL *see* biomass to liquid
building policy 161–163
Bundeswehr 23, 203
Bureau of Infrastructure, Transport and Regional Economics (BITRE) 22
Byzantium 33

CAFE *see* Corporate Average Fuel Economy
Cambridge Energy Research Associates (CERA) 15–16, 81
Campbell, Colin 9
Canada
- oil consumption 50, 133, 236
- oil sand deposits/production 36, 80–82

carbon capture and sequestration (CCS) 183
carbon dioxide emissions 179–182
cars 139–140, 142, 143–148, 216, 219
- *see also* electric vehicles

Carter, Jimmy 54, 56
Caspian Sea 35
CCS *see* carbon capture and sequestration
CERA *see* Cambridge Energy Research Associates
CFCs (chlorofluorocarbons) 184
Chávez, Hugo 37–38

Chevron 77
China
  cars 146
  oil consumption 9, 129, 132, 133, 236
chlorofluorocarbons (CFCs) 184
cities, structure of 161–163
climate change 64–65, 154–155, 178–184, 213, 221–222
coal 30, 96–97, 194
  see also liquefied coal
cognitive dissonance 241
commodities markets 65–68
compressed natural gas (CNG) 115
concentrated solar power (CSP) 121
confirmation bias 241
consumerism 169–170, 220–221
Corporate Average Fuel Economy (CAFE) standard 53
crises 52–56, 209
crude oil 1–2, 40, 75, 92
  see also oil
CSP *see* concentrated solar power
CtL *see* liquefied coal

dams 117
debt 208, 211
decrease in oil use 74–76, 139, 184–187
Deepwater Horizon accident (2010) 77
deepwater oil reserves 76–78
degrowth concept 199–200
delivery standards 239
demand for oil 50–52
  see also oil consumption
dependence on oil 4–5
derivatives 239
developing countries, oil consumption 69
Diamond, Jared 200–201
distillation of oil 3

Eagle Ford formation, Texas 84–85
economy
  collapse of 202–203
  consumerism 169–170, 220–221
  decline in oil production 193
  degrowth concept 199–200
  energy crises 140–141
  future prospects 224–227
  global economic crisis 218
  growth/oil use correlation 165–167
  levels of activity 196–197
  manufacturing 226–227
  oil and 5–7, 49–50, 164–171
  oil intensity 167–168
  peak oil 49–50
  productivity 205–208
  trust in 198–199
Egypt 43
EIA *see* Energy Information Administration
electricity 31–33, 75, 116–120, 194
electric vehicles (EVs) 119–120, 139, 219
ELM *see* Export Land Model
emissions 179–180, 181–182, 222
emotional thinking 241–242
employment 224–225
Enders, Thomas 148
energy
  crises 52–56, 140–141
  diminishing usable resources 47–48
  efficiency 168
  infrastructure 45–48, 59–61, 226
  production of 45–48, 59–61, 163–164
  security 180–181
Energy Information Administration (EIA) 8, 14–15, 22–23
energy returned on energy invested (EROEI) 27–34
  calculation 28
  coal 30
  electricity production 31–33
  fuels 32
  liquefied coal 98
  natural gas 30–31
  nuclear power 32–33
  oil 28–29
energy trap 46–47
enhanced oil recovery (EOR) 41, 73
environmental impacts
  agriculture 154–155
  cars 147
  climate change 178–184
  deepwater oil 77
  food production 154–155
  light tight oil 89–90
  liquefied coal 98–99
  oil sand 82–83
  oil shale 93
  shale oil 89–90
environmentalism 64–65, 210
EOR *see* enhanced oil recovery
ERF *see* European Union Road Federation

EROEI *see* energy returned on energy invested
ESPO pipeline 137–138
Estonia 90
ETFs *see* Exchange-traded Funds
EU *see* European Union
Europe
  agriculture 156–157
  cars 145–146
  energy sources 163
  food production 156–157
  natural gas consumption 107–108
  oil consumption 17, 69, 130–132, 236
  oil production 130
  roads 144–145
  standards of living 193
  transportation 144–145
European Union (EU) 144, 207
European Union Road Federation (ERF) 144
Exchange-traded Funds (ETFs) 66–68
exergy (diminishing usable energy) 47–48
exporters, top 12: 130
Export Land Model (ELM) 42–45, 210
ExxonMobil 87

fallacy of composition 242
FEASTA (The Foundation for the Economics of Sustainability) report 203–204
fertilizers 154, 155
financial bubbles 198–199
financial system 156, 197–199, 203–204, 230
"Finnish aviation up to 2025 – four scenarios" (Trafi) 149
Fischer–Tropsch synthesis (FT) 74, 93–96
fishing 157
flying 148–150
food production 4–5, 152–161
  Europe 156–157
  issues 159–161
  problems 155–156
  United States 157–158
forecasts, oil production 14–23, 44–45
Foss, Nicole 173
Foundation for the Economics of Sustainability (FEASTA) report 203–204
freight transport 142, 150–151, 152, 217
fuel
  efficiency 148, 149
  energy returned on energy invested 32
  prices 146–147
  replacements 194
"The Future of Oil: Geology versus Technology" (IMF) 18
future scenarios 215–232
futures markets/trading 66–68, 239, 240

gasoline 1–2, 3, 67, 134, 216
gas to liquids (GtL) 93, 114–115
Gazprom 109
GDP *see* gross domestic product
genetically modified organisms (GMOs) 158–159
geochemistry of oil 2–3
geoengineering 183
geopolitics 8–9, 175–176
Germany 23
global economic crisis 218
global reserves
  Arctic oil 78
  bitumen 80
  crude oil 16–17, 237
  deepwater oil 77–78
  kerogen 91
  shale gas 111–112
  shale oil 86–89
global security 174–178
global warming 182–183
  see also climate change
GMOs *see* genetically modified organisms
Golden Triangle 76–77
Grand Mosque, Mecca 57
Greece 202
Greenpeace 78
green revolution 153–155
Greer, John M. 194
gross domestic product (GDP) 7–8
groupthink 242
GtL *see* gas to liquids
Gulf of Mexico 77
Gustafson, Thane 135

Hansen, James 73
heavy ends (refined products) 3
Heinberg, Richard 97
highways 143
Hirsch, Robert 141
homes 227

Hormuz, Strait of 60
Hubbert, Marion King 11
hydroelectricity 117–118
hydrogen 120–123
hydropower 117

ICE *see* internal combustion engine
IEA *see* International Energy Agency
IMF *see* International Monetary Fund
importers, top 12: 130
India 119, 236
indirect oil-wealth 239
Industry Taskforce on Peak Oil and Energy Security (ITPOES) 21
INEOS Bio 95
infrastructure 45–48, 59–61, 226
institutional investors, oil futures markets 66–68
Intergovernmental Panel on Climate Change (IPCC) 213
internal combustion engine (ICE) 145
International Energy Agency (IEA) 13, 19–21, 53, 172
  oil predictions 15, 213
  World Energy Outlook 2011 135
  World Energy Outlook 2013 143
International Monetary Fund (IMF) 18–19
international trade 165
internet 205
investors, oil futures markets 66–68
involuntary degrowth 199–200
IPCC *see* Intergovernmental Panel on Climate Change
Iran 54
Iraq 175
Israel 90
ITPOES *see* Industry Taskforce on Peak Oil and Energy Security

Japan 133, 236
job market 224–225
Joint Forces Command 23

Kashagan oil field 35
Kazakhstan 35
kerogen production 90–93
Khordorkovsky, Mikhail 135
Kuwait 61
Kuzniki, Steven 79

labor unions 220
Laherrère, Jean 17–18
Libya 35, 36
lifestyle 223
light ends (refined products) 3
light tight oil (LTO) 83–90
  depletion of resources 88
  environmental impacts 89–90
  global resources 86–89
  production 83–86
Limnell, Jarno 177
liquefied biomass (BtL) 93
liquefied coal (CtL) 93, 96–99
liquefied natural gas (GtL) 93, 114–115
liquid petroleum gas (LPG) 105, 106
location of oil reserves 34–35
Loppacher, Laura 111
LTO *see* light tight oil

MacKay, David 142
Malkki, Janne 177
manufacturing 226–227
marginal utility concept 167
markets 171–174, 197, 238–241
media 212–214
medium distillates (refined products) 3
methane emissions 222
Metis Risk Consulting report 203–204
Mexico, Gulf of 77
military action 44, 176–178

nationalism, resource 36–37
national oil companies (NOCs) 37
National Security Study Memorandum 200 (1974) 42
natural fallacy 241–242
natural gas (NG)
  compressed natural gas 115
  conventional natural gas 106–107
  energy returned on energy invested 30–31
  European consumption 107–108
  Russian production 107, 108–109
  transportation 105, 115–116
  US production 110–111
natural gas liquid (NGL) 105–106, 113–114
news media 212–214
NG *see* natural gas
NGL *see* natural gas liquid
NOCs *see* national oil companies
nuclear power 32–33, 44, 45

Obama, Barack 68
ocean transportation 150–151
OECD *see* Organization for Economic Cooperation and Development

oil
units and terms 235–236
uses of 1
oil consumption
1970–2010: 133
aviation 148–149
Canada 133
China 132, 133
demand increase 50–52
developing countries 69
Europe 17, 69, 130–132, 236
global 128, 236–237
gross domestic product 7–8
Japan 133
OECD countries 192–193
reduction in 74–76
Russia 133, 137
top 12 exporters/importers, 2012: 130
United States 17, 132–134
oil exchanges 240
oil industry, birthplace 1
oil prices 40, 192, 195, 211–212
commodities markets 65–68
highest price 49–50
maximum price 69
oil crises 52–56
permanently expensive oil prospect 51–52
speculators 66
volatility of 68–69
oil-producing nations
OPEC 52, 54, 61–62
political instability 56–58
oil production 15, 129, 130, 237, 238
decline in 73–76, 193
efficiency of 39–42
energy returned on energy invested 28–29
Europe 130
forecasts 14–23, 44–45
Russia 134–138
technology 39
United States 132
oil reserves
Arctic oil 78–79
backdating added reserves 63
deepwater oil 76–78
global 8–9, 77–78, 86–89
Golden Triangle 76–77
location 34–35
paper oil reserves 61–63
political oil reserves 61–63
quality 34–35
Russia 136–138
strategic resources 8–9
see also global reserves
oil sands 36, 79–83
oil scarcity, policies 171–174
oil shale 90–93
oil substitutes 74
OPEC *see* Organization of the Petroleum Exporting Countries
organic farming 158
Organization for Economic Cooperation and Development (OECD) countries 19, 138–140, 192–193
Organization of the Petroleum Exporting Countries (OPEC) 52, 54, 61–62
see also oil-producing nations
ozone layer 184

paper oil reserves 61–63
payments 240
PDVSA (Venezuelan national oil company) 37–38
peak oil
climate change 178–184
dynamics of 26–72
economic limitations and motivations 49–50
effects of 194–196
forecasts 10, 12–23
history of theory 11–12
media 212–214
officials and 210–212
oil reserves 61–63
overview 9–23
perspective 48
politics 214–215
predictions 10, 12–23
public debate 211–214
variables 10–11
Pearce, Fred 96
peat 222
Peeking at Peak Oil (Aleklett) 137, 152
Persian Gulf 60
petroleum gases 3
pipelines 59, 137–138
plastics 4
policy suggestions 223–232
political oil reserves 61–63
politics 56–58, 208–209, 214–215, 223–224
population 152
ports 60–61

Powers, Bill 109
predictions, oil production 14–23, 44–45
price of oil *see* oil prices
PricewaterhouseCoopers (PwC) report, shale oil 87–88
primary economy 196, 226
prisoners 219
productivity 205–208
public sector 218, 229–230
Putilov, Aleksander 38
Putin, Vladimir 135
pyramid schemes 209–210

quality of oil 34–36

railways 142, 145, 151
rational thinking, difficulty of 241–242
Reagan, Ronald 54
reduction in oil use 74–76, 139, 184–187
refineries 3–4, 59–61
regulation 64–65
resource nationalism 36–37
ride-sharing services 231–232
risk
  biofuels 103–104
  deepwater oil production 77
  see also environmental impacts
roads 143, 144–145
Rosneft 38–39, 87, 135, 138
Royal Dutch Shell 63
Rubin, Jeff 5
rural areas 173, 220
Russia
  influence of 229
  natural gas production 107, 108–109
  oil consumption 133, 137, 236
  oil exports 137
  oil production 134–138
  oil reserves 34, 41, 136–138
  see also Rosneft; Soviet Union

SAGD *see* Steam Assisted Gravity Drainage
Sakhalin Island, Russia 41
Saudi Arabia 35, 36, 40–41, 57–58
  economy 43
  nuclear power 44
  oil consumption 236
schools 227
secondary economy 196
second homes 227
security 174–178, 180–181, 230
services sector 228
Shah, R.R. 99
shale gas 109–112
shale oil 83–90
  depletion of resources 88
  environmental impacts 89–90
  global resources 86–89
  production 83–86
Shell 63
shipping 150–151
Siberian oilfields 34, 137
social security systems 201, 218
society
  collapse of 200–203
  EROEI threshold 29
  fragility of 54–56
  net energy for 29
  oil scarcity policies 171–174
  polarization of 219–220
  simplification of 201–202
  structure of 161–163
soil 155, 156
solar power 121
Soviet Union 41, 202–203
  see also Russia
Spain 204
speculators 66
spot-prices 239
standards of living 193
Steam Assisted Gravity Drainage (SAGD) 81, 82
strategic resources 8–9
stripper wells 41
suburbs 143
Sweden 22

Tainter, Joseph A. 200, 201
Taleb, Nassim Nicholas 26, 214
technology 206
terminology 235–236
tertiary economy 196
thermolysis 121
tourism 229
trade 165
Trafi (Finnish Transport Safety Agency) 149
tragedy of the commons 241
transfers 240
Transition Movement 212
transportation 141–152
  aviation 148–150
  cars 142, 143–148
  demand for oil 4
  electric vehicles 119–120

electricity 75, 116, 119–120
Europe 144–145
future prospects 227–228
infrastructure 144–145
logistics 141–152
natural gas 105, 115–116
of oil 34–35, 60–61, 175
oil use 128–129, 139, 173, 216–218
railways 145, 151
ride-sharing services 231–232
shipping 150–151
without oil 105
trucks, Caterpillar 797 81
trust 198–199

UGES *see* Uppsala Global Energy Systems
unions 220
United Kingdom 21, 236
United States
agriculture 157–158
cars 146
Corporate Average Fuel Economy standard 53
currency 168–169
Energy Information Administration 8, 14–15, 22–23
EROEI of oil production 29
food production 157–158
gasoline consumption 134
Joint Forces Command 23
military power of 58
natural gas production 110–111
oil consumption 129, 132–134, 236
oil production 132
shale oil production 84–85
suburban sprawl 143
units and terms 235–236
UPM biorefinery 94
Uppsala Global Energy Systems (UGES) 22
Ural-Volga oilfields 137
uses of oil 4–5

Vankor oilfield 138
Venezuela 37–38, 61
VTT Technical Research Center, Finland 95

warfare 44, 176–178
welfare society 205–208
Wilkerson, Lawrence 59, 61
wind power 118
World Energy Outlook (IEA) reports 13, 135, 143
world history, oil's role 174–175

Yergin, Daniel 15–16

zoning 161–163, 212